CAMBRIDGE LIBRARY COLLECTION

Books of enduring scholarly value

Physical Sciences

From ancient times, humans have tried to understand the workings of the world around them. The roots of modern physical science go back to the very earliest mechanical devices such as levers and rollers, the mixing of paints and dyes, and the importance of the heavenly bodies in early religious observance and navigation. The physical sciences as we know them today began to emerge as independent academic subjects during the early modern period, in the work of Newton and other 'natural philosophers', and numerous sub-disciplines developed during the centuries that followed. This part of the Cambridge Library Collection is devoted to landmark publications in this area which will be of interest to historians of science concerned with individual scientists, particular discoveries, and advances in scientific method, or with the establishment and development of scientific institutions around the world.

The Life of Sir Humphry Davy

Sir Humphry Davy (1778–1829) was a hugely influential chemist, inventor, and public lecturer who is recognised as one of the first professional scientists. He was apprenticed to an apothecary in 1795, which formed his introduction to chemical experiments. A chance meeting with Davis Giddy in 1798 introduced Davy into the wider scientific community, and in 1800 he was invited to a post at the Royal Institution, where he lectured to great acclaim. These volumes, first published in 1831, contain Davy's official biography. Researched and written by John Ayrton Paris, the work describes in detail Davy's life and his scientific studies. Organised chronologically with excerpts from his private correspondence, Davy's early life and his experiments and lectures at the Royal Institution and his Presidency of the Royal Society between 1820 and 1827 are explored in vivid detail. Volume 2 describes his life and work between 1812 and 1829.

Cambridge University Press has long been a pioneer in the reissuing of out-of-print titles from its own backlist, producing digital reprints of books that are still sought after by scholars and students but could not be reprinted economically using traditional technology. The Cambridge Library Collection extends this activity to a wider range of books which are still of importance to researchers and professionals, either for the source material they contain, or as landmarks in the history of their academic discipline.

Drawing from the world-renowned collections in the Cambridge University Library, and guided by the advice of experts in each subject area, Cambridge University Press is using state-of-the-art scanning machines in its own Printing House to capture the content of each book selected for inclusion. The files are processed to give a consistently clear, crisp image, and the books finished to the high quality standard for which the Press is recognised around the world. The latest print-on-demand technology ensures that the books will remain available indefinitely, and that orders for single or multiple copies can quickly be supplied.

The Cambridge Library Collection will bring back to life books of enduring scholarly value (including out-of-copyright works originally issued by other publishers) across a wide range of disciplines in the humanities and social sciences and in science and technology.

The Life of Sir Humphry Davy

VOLUME 2

JOHN AYRTON PARIS

CAMBRIDGE UNIVERSITY PRESS

Cambridge, New York, Melbourne, Madrid, Cape Town,
Singapore, São Paolo, Delhi, Tokyo, Mexico City

Published in the United States of America by Cambridge University Press, New York

www.cambridge.org
Information on this title: www.cambridge.org/9781108073196

This edition first published 1831
This digitally printed version 2011

ISBN 978-1-108-07319-6 Paperback

THE LIFE

OF

SIR HUMPHRY DAVY,

BART. LL.D.

LATE PRESIDENT OF THE ROYAL SOCIETY, FOREIGN ASSOCIATE
OF THE ROYAL INSTITUTE OF FRANCE,
&c. &c. &c.

BY

JOHN AYRTON PARIS, M.D. CANTAB. F.R.S. &c.

FELLOW OF THE ROYAL COLLEGE OF PHYSICIANS.

IN TWO VOLUMES.

VOL. II.

LONDON:
HENRY COLBURN AND RICHARD BENTLEY,
NEW BURLINGTON STREET.
M DCCC XXXI.

CONTENTS.

CHAPTER X.

CHAPTER XI.

CHAPTER XII.

CHAPTER XIII.

THE LIFE

OF

SIR HUMPHRY DAVY,

BART. &c. &c.

CHAPTER X.

Mr. Faraday's introduction to Sir H. Davy.—A renewed correspondence on the subject of the Gunpowder Manufactory.—Davy obtains permission from Napoleon to visit the Continent.—He embarks in a Cartel from Plymouth.—Is arrested at Morlaix.—Arrives at Paris.—Visits the Louvre.—His extraordinary conduct upon that occasion.—Inspects the Colossal Elephant, and is introduced to M. Alavair, its architect.—The discovery of the dungeons of the Bastile.—Davy's interesting letter to M. Alavair.—He attends a meeting of the Institute.—Is visited by all the principal *savans* of Paris.—The adventure which befell Lady Davy in the Thuilleries' Garden.—Anniversary dinner of the Philomatic Society.—The junior Chemists of France invite Davy to a splendid entertainment.—How far Davy is entitled to be considered the discoverer of the true nature of Iodine.—Napoleon's unlucky experiment with the Voltaic battery.—Davy is presented to the Empress Josephine.—An account of the Court ceremony at Malmaison.—Remarks on the conduct of Davy during his visit to Paris.—He quits the capital of France, and proceeds, by way of Lyons, to Montpellier.—Is assisted in experiments on sea-weed by M. Berard.—Crosses the Alps.—Arrives at Genoa.—Institutes experiments on the Torpedo.—Visits Florence, and accomplishes the combustion of the diamond, by the great lens in the cabinet of Natural History.—Experiments on Iodine.—He examines the colours used by the

Ancients.—Visits all the celebrated Philosophers of Italy and Switzerland, with whom he works in their laboratories.—Returns to England.

It is said of Bergman, that he considered the greatest of his discoveries to have been the discovery of Scheele.* Amongst the numerous services conferred upon Science by Sir Humphry Davy, we must not pass unnoticed that kind and generous patronage which first raised Mr. Faraday from obscurity, and gave to the chemical world a philosopher capable of pursuing that brilliant path of enquiry which the genius of his master had so successfully explored.

The circumstances which first led Mr. Faraday to the study of chemistry, and by which he became connected with the Royal Institution, were communicated to me, by himself, in the following letter.

TO J. A. PARIS, M. D.

MY DEAR SIR, Royal Institution, Dec. 23, 1829.

You asked me to give you an account of my first introduction to Sir H. Davy, which I am very happy to do, as I think the circumstances will bear testimony to his goodness of heart.

When I was a bookseller's apprentice, I was very fond of experiment, and very averse to trade. It happened that a gentleman, a member of the Royal Institution, took me to hear some of Sir H. Davy's last lectures in Albemarle Street. I took notes, and afterwards wrote them out more fairly in a quarto volume.

* See Note at page 42, vol. i.

My desire to escape from trade, which I thought vicious and selfish, and to enter into the service of Science, which I imagined made its pursuers amiable and liberal, induced me at last to take the bold and simple step of writing to Sir H. Davy, expressing my wishes, and a hope that, if an opportunity came in his way, he would favour my views; at the same time I sent the notes I had taken at his lectures.

The answer, which makes all the point of my communication, I send you in the original, requesting you to take great care of it, and to let me have it back, for you may imagine how much I value it.

You will observe that this took place at the end of the year 1812, and early in 1813 he requested to see me, and told me of the situation of assistant in the Laboratory of the Royal Institution, then just vacant.

At the same time that he thus gratified my desires as to scientific employment, he still advised me not to give up the prospects I had before me, telling me that Science was a harsh mistress; and, in a pecuniary point of view, but poorly rewarding those who devoted themselves to her service. He smiled at my notion of the superior moral feelings of philosophic men, and said he would leave me to the experience of a few years to set me right on that matter.

Finally, through his good efforts I went to the Royal Institution early in March of 1813, as assistant in the Laboratory; and in October of the same year, went with him abroad as his assistant in experiments and in writing. I returned with him in April 1815, resumed my station in the Royal Institution, and have, as you know, ever since remained there.

I am, dear Sir, very truly yours,
M. FARADAY.

The following is a note of Sir H. Davy, alluded to in Mr. Faraday's letter:

TO MR. FARADAY.

SIR, December 24, 1812.

I AM far from displeased with the proof you have given me of your confidence, and which displays great zeal, power of memory, and attention. I am obliged to go out of town, and shall not be settled in town till the end of January: I will then see you at any time you wish.

It would gratify me to be of any service to you. I wish it may be in my power.

I am, Sir, your obedient humble servant,
H. DAVY.

I must now recall the reader's attention to the affair of the gunpowder manufactory, to which some allusion has been already made. It is far from my wish to intrude upon the public any account of a private transaction; but the circumstances to which I must refer are already well known, and I believe, moreover, that they have been the subject of misrepresentation.

The letters I shall introduce appear to me highly interesting; and by the warmth of feeling with which they repel the bare suspicion of his prostituting science to the acquisition of wealth, to develope a feature in his character too important to be omitted in a memoir of his life.

From the following letter, it would appear that Davy's alarms, with respect to his responsibilities, were first awakened by a sight of the labels, in which his name was introduced.*

TO JOHN GEORGE CHILDREN, ESQ.

MY DEAR CHILDREN, Rokeby, July — 1813.

I AM very sorry you did not come to Cobham, as the party was very pleasant.

Your apparatus was magnificent, worthy an Imperial Institute: there were some swine however for the pearls; at least, there was one,—you cannot suppose I mean any other than ——

I have been much disturbed and vexed by enquiries respecting the price of *my* gunpowder, which from the labels I find is supposed to be *sold* by me. These labels must be altered, so as to put in a clear point my relations to the manufacture; and it must be understood by the public that I have given my gratuitous assistance and advice only.

I have written to Mr. Burton by post, giving two forms. I shall do you more good if these are adopted than I can now; and I wish them to be adopted

* I am here bound to state, from a careful examination of all the original documents, that his name was introduced in the very words which he suggested, and which I have at this moment before me in his own handwriting:—so differently, however, does the same sentence strike the eye in print and in manuscript, that an author frequently does not recognise his own composition.

speedily, as it may otherwise get abroad that I have nothing to do with the powder, and that my name is used in a manner which does not meet my approbation.

In the labels in the windows, it should not be *under my directions*, for this implies that I am a superintendent in the manufactory; but it should be —" RAMHURST GUNPOWDER, manufactured by Messrs. B. C. and Co. In the composition of this powder, the proprietors have been assisted by the advice and assistance of Sir H. Davy."

A fair statement will do the manufacture good. Misapprehension will do it much harm.

I am now at Rokeby; we shall be in a few days at Braham Castle, Lord Mackenzie's, near Dingwall, where we shall stay for a week. After that we shall go to the Marquis of Stafford's, Dunrobin, near Goldspie.

I am, my dear Children,

Very truly and affectionately yours,

H. DAVY.

TO THE SAME.

MY DEAR CHILDREN, Edinburgh, July 22.

I WROTE to you from Rokeby. I expressed my feelings respecting the gunpowder. I have been in extreme harass and anxiety from the idea of the use of my name, without the proper explanation, and I certainly expected that no use would have been made of it without my sanction. I never saw the label for the canister till it came to me upon one of them, and I immediately expressed that I was not satisfied with it.

I told Mr. Burton expressly, that in all cases in which my name was used it must be in my own way. He is now at the head of your firm; but it is to *you*, and not to *him*, that I have given, and shall give my assistance.

Every feeling of friendship and affection prompts my wishes to be useful to you; I have not the same relations to Mr. Burton.

I am very sorry to give you any trouble on this business, but I am sure you cannot wish me to remain in a state of anxiety; and all the friends with whom I have consulted think it absolutely necessary for my reputation, that, when my name is used, a clear statement should be given of the true nature of the connexion.

I think it will be more useful to you, and increase your influence and power in the partnership, if my assistance is stated as given to you, and to you only—in this way: "RAMHURST GUNPOWDER, manufactured by Messrs. Burton, Children, and Co. after an improved process, founded upon experiments and investigations made by Sir H. Davy, and communicated by him to Mr. J. G. Children, under whose immediate superintendence the gunpowder is made."

I have fully made up my mind on this matter: and if you approve of the above form, I will state it to be the only one to which I will consent.

If the gunpowder is called Sir H. Davy's powder, it must be stated in all cases where my name is used, that it is so called in honour of my discoveries in chemistry, and because I have given my gratuitous assistance in making the experiments and investigations on which the process is founded.

I have resolved to make no profit of any thing connected with science. I devote my life to the public in future, and I must have it clearly understood, that I have no views of profit in any thing I do. I am, my dear Children,

Very affectionately yours,

H. DAVY.

In subsequent letters, which it is not necessary to publish, Davy dwells upon the necessity of his engagements as a partner being legally cancelled, as he cannot endure the idea of his philosophical repose and usefulness being disturbed by the cares of business, or the trouble of litigation.

It is scarcely necessary to add, that all the parties concerned in this transaction most readily and cheerfully met Davy's wishes, all erroneous impressions were effaced, and the affair was adjusted amicably and satisfactorily; and he prepared to quit England with a mind relieved from all the fears and anxieties which had so unfortunately oppressed it.

After the Emperor of the French had sternly refused his passport to several of the most illustrious noblemen of England, it was scarcely to be expected that Sir H. Davy would have been allowed to travel through France, in order to visit the extinct volcanoes in Auvergne, and afterwards to examine that which was in a state of activity at Naples.

No sooner, however, had the discovery of the decomposition of the alkalies and earths, and its probable bearings upon the philosophy of volcanic action, been represented by the Imperial Institute to Napoleon, than, with a liberality worthy of the liberator of Dolomieu, and consistent with his well-known patronage of science, he immediately and unconditionally extended the required indulgence.

In consequence of this permission, Sir Humphry and Lady Davy, the former accompanied by Mr. Faraday as secretary and chemical assistant, and the latter by her own waiting-maid, quitted London on the 13th of October 1813, and proceeded to Plymouth; at which port they immediately embarked in a cartel for Morlaix in Brittany.

On landing in France, they were instantly arrested by the local authorities of the town, who very reasonably questioned the authenticity of their passports, believing it impossible that a party of English should, under any circumstances, have obtained permission to travel over the Continent, at a time when the only English in France were detained as prisoners. They were accordingly compelled to remain during a period of six or seven days at the town of Morlaix, until the necessary instructions could be received from Paris. As soon, however, as a satisfactory answer was returned, they were set at liberty; and they reached the French capital on the evening of the 27th of the same month.

Shortly after his arrival, Davy called upon his old friend and associate Mr. Underwood, who, although one of the *détenus*, had during the whole war enjoyed the indulgence of residing in the capital.

The expected arrival of Davy had been a subject of conversation with the French *savans* for more than a month. Amongst those who were loudest in his praises, was M. Ampère, who had for several years frequently expressed his opinion that Davy was the greatest chemist that had ever appeared. Whether this flattering circumstance had been communicated to the English philosopher I have no means of ascertaining; but Mr. Underwood informs me that the very first wish that Davy expressed was to be introduced to this gentleman, whom he considered as the only chemist in Paris who had duly appreciated the value of his discoveries; an opinion which he afterwards took no care to conceal, and which occasioned amongst the *savans* much surprise, and some dissatisfaction. M. Ampère, at the time of Davy's arrival, was spending the summer at a place a few miles from Paris, in consequence of which the introduction so much desired was necessarily delayed.

On the 30th he was conducted by Mr. Underwood to the Louvre. The English philosopher walked with a rapid step along the gallery, and, to the great astonishment and mortification of his friend and *cicerone*, did not direct his attention to a single painting; the only exclamation of surprise that escaped him was—"What an extraordinary collection of fine frames!"—On arriving opposite to Raphael's picture of the Transfiguration, Mr. Underwood could no longer suppress his surprise, and in a tone of enthusiasm he directed the attention of the philosopher to that most sublime production of art, and the chef d'œuvre of the collection. Davy's reply was as laconic as it was chilling—"Indeed, I am glad I have seen it;" and then hurried forward, as if he were desirous of escaping from any critical remarks upon its excellencies.

They afterwards descended to view the statues in the lower apartments: here he displayed the same frigid indifference towards the higher works of art. A spectator of the scene might have well imagined that some mighty spell was in operation, by which the order of nature had been reversed:— while the marble glowed with more than human passion, the living man was colder than stone! The apathy, the total want of feeling he betrayed on having his attention directed to the Apollo Belvidere, the Laocoon, and the Venus de Medicis, was as inexplicable as it was provoking; but an exclamation of the most vivid surprise escaped him at the sight of an Antinous, treated in the Egyptian style, and sculptured in *Alabaster.** — "Gracious powers," said he, "what a beautiful stalactite!"

What a strange—what a discordant anomaly in the construction of the human mind do these anecdotes unfold! We have here presented to us a philosopher, who, with the glowing fancy of a poet, is insensible to the divine beauties of the sister arts! Let the metaphysician, if he can, unravel the mystery,—the biographer has only to observe that the Muses could never have danced in chorus at his birth.

On the following morning, Mr. Underwood accompanied him to the Jardin des Plantes, and presented him to the venerable Vauquelin, who was the first scientific man he had seen in Paris. On their

* The celebrated Italian antiquary Visconti has so denominated it.

return they inspected the colossal Elephant which was intended to form a part of the fountain then erecting on the site of the Bastile. Davy appeared to be more delighted with this stupendous work than with any object he saw in Paris: to its architect, M. Alavair, he formed an immediate attachment. It has been observed that, during his residence in this city, his likes and dislikes to particular persons were violent, and that they were, apparently, not directed by any principle, but were the effect of sudden impulse.

In the course of removing the foundations, and in digging the canal, the subterranean dungeons of the Bastile were discovered; they were eight in number, and were called *Les Oubliettes.* As they were under the level of the ditch of the fortress, any attempt to escape from them by piercing the wall, must have inevitably drowned the unhappy prisoner together with all those who inhabited the contiguous cells; one of which was discovered with the entrance walled up. Upon demolishing this wall there appeared the skeleton of the last wretched person who had been thus entombed. In all these discoveries Davy took the warmest interest.

Upon the construction of the Elephant, he wrote a letter to M. Alavair, to which I am desirous of directing the attention of my scientific readers. It derives its peculiar interest from the opinion which he at that period entertained upon the subject of the excitement of Voltaic action by the contact of different metals.

TO M. ANTOINE ALAVAIR.

SIR, November 1813.

It will give me much pleasure if I can repay your civility to me by offering any hints that may be useful in the execution of the magnificent work constructing under your directions.

Ten parts of copper to one of tin is an excellent composition for a work upon a great scale, nor do I believe any proportions can be better.

There is no fear of any decay in the armatures, if they can be preserved from the contact of moisture; but if exposed to air and moisture, the presence of the bronze will materially assist their decay. Wherever the iron is exposed to air, it should, if possible, be covered with a thin layer of bronze. When the iron touches the foundation of *lead*, it should in like manner be covered either by lead or bronze. A contact between metals has no effect of corrosion, unless a Voltaic circle is formed with moisture, and then the most oxidable metal corrodes; and iron corrodes rapidly both with lead and bronze.

The cement which will probably be found the most durable will be lime, fine sand, and scoria of iron. The materials should be very fine and intimately mixed. The ancients always made their cements for great works some months before they were used. I have the honour to be, Sir, with much consideration,

Your obedient humble servant,

H. Davy.

Davy took up his abode at the Hotel des Princes, Rue Richelieu; whither the principal *savans* of

Paris hastened to pay their respects; which they did with an alacrity and cheerfulness equalled only by the courtesy of manner with which they expressed their congratulations.

On the 2nd of November, Davy attended the First Class of the Institute, and was placed on the right hand of the President, who on taking the chair announced to the meeting that it was honoured by the presence of "Le Chevalier Davy."

While Davy was at the meeting of the Institute, a curious adventure occurred to Lady Davy, the relation of which, by showing the state of surveillance in which the citizens of Paris were held at that period, will enable us to appreciate the extent of the obligation conferred upon Sir Humphry by the Emperor.

Her Ladyship, attended by her maid, had walked into the Thuilleries' Garden. She wore a very small hat, of a simple cockle-shell form, such as was fashionable at that time in London; while the Parisian ladies wore bonnets of most voluminous dimensions. It happened to be a saint's day, on which, the shops being closed, the citizens repaired in crowds to the garden. On seeing the diminutive bonnet of Lady Davy, the Parisians felt little less surprise than did the inhabitants of Brobdignag on beholding the hat of Gulliver; and a crowd of persons soon assembled around the unknown exotic; in consequence of which, one of the inspectors of the garden immediately presented himself, and informed her Ladyship that no cause of '*rassemblement*' could be suffered, and therefore requested her to retire. Some officers of the Imperial Guard, to whom she appealed, replied, that, however much they might regret the circumstance, they were unable to afford her any redress, as the order was peremptory. She then requested that they would conduct her to her carriage; an officer immediately offered his arm, but the crowd had by this time so greatly increased, that it became necessary to send for a corporal's guard; and the party quitted the garden, surrounded by fixed bayonets.

November 3rd, Humboldt and Gay-Lussac paid their first visit of compliment to Davy.

5th.—M. Ampère, who came to Paris expressly for the occasion, was introduced to Davy by Mr. Underwood, and the two philosophers appeared equally delighted with each other. Some years afterwards, however, this feeling of friendly regard, on the part of Davy, was turned into one of bitter aversion, in consequence, as it has been supposed, of certain perfidious insinuations, by which some of the *savans*, instigated by feelings of jealousy, had contrived to prejudice his mind; and which even led him to exert all his efforts to oppose the election of Ampère as a foreign member of the Royal Society.

After Ampère's visit, Mr. Underwood accompanied Sir H. and his lady to the Imperial Library in the Rue Richelieu, and afterwards to the Cathedral of Notre Dame, where they inspected the crown and imperial regalia. The splendour of the coronation mantle of Napoleon may be imagined, when it is stated that its weight exceeded eighty pounds, and that it was lined with the skins of six thousand ermines.

6th.—They visited the Museum of French Monuments, in the Rue des Petits Augustines, which contained the tombs and sculptured ornaments preserved from the churches that were demolished during the Revolution. This interesting collection, shortly after the restoration of the Bourbons, was dispersed. It is a singular fact, that Davy expressed more admiration at this inferior exhibition of art, than he did at that of the Greek and Roman statues in the Museum of the Louvre. Whether his taste had been vitiated by the inspection of less perfect models in his earlier days, is a question which I shall leave more competent judges to decide.

10th. — They dined with Count Rumford at Auteuil, who showed his laboratory to Davy: this was exactly eight months before the poor broken-hearted Count sank into the grave, the victim of domestic torment, and of the persecutions of the French *savans*, instigated by his wife, the widow of the celebrated Lavoisier.

13th.—The anniversary dinner of the Philomatic Society took place on this day, at a restaurateur's in the Rue St. Honoré; M. Dumeril in the chair. Although it was very unusual to invite any stranger upon this occasion, Sir H. Davy and his English friend were requested to favour the company by their presence. Thirty-three members were in attendance, amongst whom were Ampère, Brogniart, Cuvier, Chevreuil, Dulong, Gay-Lussac, Humboldt, Thénard, &c.

At this dinner various complimentary toasts were proposed: and first, the Royal Society of London, for which Davy having returned thanks, gave the Imperial Institute. The Linnæan Society of London, and the Royal Society of Berlin, were given in succession. But the circumstance which evinced the greatest feeling and delicacy towards their English guest, was the company's declining to drink the health of the Emperor. It placed their personal safety even in some jeopardy; and not a little apprehension was afterwards felt as to how far Napoleon might resent such a mark of disrespect, for seven-eighths of the members present were placemen.

November 17th.—Mr. Underwood states that on this day he met Humboldt at dinner at Davy's hotel; and he adds—"I do not know whether you are aware that Sir Humphry had a superstitious dislike at seeing a knife and fork placed crosswise on a plate at dinner, or upon any other occasion; but I can assure you that such was the fact; and when it occurred in the company of his intimate friends, he always requested that they might be displaced; whenever this could not be done, he was evidently very uncomfortable."*

At about this period, but I am unable to ascertain the particular day, the junior Chemists, with Thénard as their leader, gave Davy a sumptuous dinner at one of the most celebrated restaurateurs in

* I repeat this as I received it: from my own personal knowledge I can neither confirm nor refute it; although I am inclined to believe that Davy was tinged with a degree of superstitious feeling, or a certain undefined species of credulity, which shelters itself under the acknowledged inadequacy of human reason to connect causes with effects.

Paris. The following persons formed a committee for that purpose: Gay-Lussac, Thénard, Dulong, Chevreul, Laugier, Robiquet, and Clement. As it was by the chemists only that this dinner was given, neither Arago nor Ampère was included; but Berthollet, Chaptal, and Vauquelin were invited.

On the morning of the 23rd of November,* M. Ampère called upon Davy, and placed in his hands a small portion of a substance which he had received from M. Clement; and, although it had been in the possession of the French chemists for more than twelve months, so entirely ignorant were they of its true nature and composition, that it was constantly spoken of amongst themselves as X, the *unknown* body.

How far the suggestions of Davy led to the discovery of the chemical nature of this interesting substance, which has been since distinguished by the name of *Iodine*, is a question which has given rise to much discussion on the Continent. It has been moreover questioned, how far the love of science, and the fervour of emulation, can justify the interference which Davy is said to have displayed upon the occasion. He is accused of having unfairly taken the subject out of the hands of those who were engaged in its investigation, and to have anticipated their results.

As his biographer, I feel that it is not only due to the character of Davy, but essential to the history of Science, that these questions should be im-

* The date of this event is important; and Mr. Faraday, in referring to his Journal, finds it to be correctly stated.

partially examined; and I have spared no pains in collecting facts for their elucidation. Mr. Underwood, who was in the constant habit of associating with the parties concerned in the enquiry, has furnished me with some important particulars, and his testimony is fortified by published documents.

The substance under dispute was accidentally discovered by M. Courtois, a manufacturer of saltpetre at Paris, but kept secret by him for several years; at length, however, he communicated it to M. Clement, who made several experiments on it, but without any favourable result. On the 23rd of August 1813, Clement exhibited to Mr. Underwood the beautiful experiment of raising it into a violet-coloured vapour, and that gentleman assures me that this was the only peculiar property which had at that time been recognised as distinguishing it. A few days previous to this event, M. Ampère had received a specimen of the substance, which he had carefully folded up in paper, and deposited in his pocket, but on arriving at home, and opening the packet, he was surprised to find that his treasure had vanished. Clement, however, furnished him with another supply, and it was this parcel that Ampère transferred into the hands of Davy; and "for which," says Mr. Underwood, "he told me a few days ago, that Thénard and Gay-Lussac were extremely angry with him."

The first opinion which the French chemists entertained respecting the nature of Iodine, was that it was either a compound of muriatic acid, or of chlorine, since it formed with silver what appeared to be a muriate, or a chloride of that metal; but

Davy at once observed that the substance so produced blackened too quickly in the sun to justify that opinion. He, however, determined to submit it to a more rigorous examination; and during the latter part of November he worked upon it at his hotel with his own apparatus, and on the 3rd of December in the laboratory of M. Chevreul, at the Jardin des Plantes, with whom, it may be stated in passing, he perhaps formed a stricter intimacy than with any other chemist during his sojourn in Paris. Chevreul, however, be it known, was a brother of the angle; and I understand that he still preserves, as sacred trophies, some artificial flies with which Davy had supplied him.

Having pointed out the channel through which Iodine first fell into the hands of Davy, let us pursue its history. The first public notice of its existence was read by Clement at the Institute, on the 29th of November 1813. At the meeting of the 6th of December, Gay-Lussac, who had only received some 'X' a few days previous to this date, presented a short note, in which he gave the name of *Iode* to the body, and threw out a hint as to its great analogy to chlorine, while he stated that two hypotheses might be formed as to its nature, that it might be considered as a simple substance, or as a compound of oxygen. On the 13th of the same month, a letter addressed to M. le Chevalier Cuvier, and dated December 11th, was read from Davy to the Institute, in which he offered a general view of its chemical nature and relations;* and on the 20th

* See *Annales de Chimie*, tome 88, p. 322. It appears from Mr. Faraday's Journal, that he worked upon Iodine with a borrowed

of January 1814, he communicated to the Royal Society of London, a long and elaborate paper, dated Paris, December 10, 1813, and entitled, " Some Experiments and Observations on a new Substance, which becomes a violet-coloured gas by Heat." In this paper, while the author assigned to Gay-Lussac all the credit to which his communication of the 6th of December may be supposed to entitle him, he evidently felt that some explanation was due to the chemical world for his having pursued the enquiry. " M. Gay-Lussac," he observes, " is still engaged in experiments on this subject, and from his activity and great sagacity, a complete chemical history of it may be anticipated. But as the mode of procuring the substance is now known to the chemical world in general, and as the combinations and agencies of it offer an extensive field for enquiry, and will probably occupy the attention of many persons; and as the investigation of it is not pursued by the discoverer himself, nor particularly by the gentlemen to whom it was first communicated, I shall not hesitate to lay before the Royal Society an account of the investigations I have made upon it; and I do this with the less scruple, as my particular manner of viewing the phenomena has led me to some new results, which probably will not be considered by the Society as without interest in their relation to the general theory of chemistry, and in their possible application to some of the useful arts."

Voltaic pile, at his hotel, on the morning of the 11th; and the results of his experiments are described at the conclusion of the above letter.

It was not until August 1814, that Gay-Lussac read his paper on the subject, which was subsequently published in the *Annales de Chimie.*

After the above short, but I trust honest statement, can any reasonable doubt exist, that, if Davy had not visited Paris, Iodine would have remained at the end of the year 1814, as it had been for two preceding years—the unknown X?

In a communication published in the first volume of the Royal Institution Journal, Davy offers the following observations upon this subject: "With regard to Iodine, the first I had of it was from M. Ampère, who, before I had seen the substance, supposed that it might contain a new supporter of combustion.

"Who had most share in developing the chemical history of that body, must be determined by a review of the papers that have been published upon it, and by an examination of their respective dates. When M. Clement showed Iodine to me, he believed that the hydriodic acid was muriatic acid; and M. Gay-Lussac, after his early experiments, made originally with M. Clement, formed the same opinion, and *maintained* it, when I *first* stated to him my belief that it was a new and peculiar acid, and that Iodine was a substance analogous in its chemical relations to Chlorine."

I was very desirous of ascertaining the feeling which at present prevails amongst the French chemists upon this subject; and I therefore requested Mr. Underwood to make such enquiries as might elicit the required information. In a letter from that gentleman, dated "Paris, August 22, 1830," he says, " Though Thénard and Gay-Lussac retain great bitterness of feeling towards Davy, on account of the affair of Iodine, Chevreul and Ampère are still, as they ever were, of opinion, that such a feeling has its origin in a misconception; that what Davy did, was from the honest desire of promoting science, and not from any wish to detract from the merit of the French chemists."

During his visit to Paris, Davy was not introduced to the Emperor. Lady Davy observed to me, that, although Sir Humphry felt justly grateful for the indulgence granted to him as a Philosopher, he never, for a moment, forgot the duty he owed his country as a Patriot; and that he objected to attend the levee of her bitterest enemy. On the other hand, it is said that Napoleon never expressed any wish to receive the English chemist; and those who seek in the depths for that which floats upon the surface, have racked their imaginations in order to discover the source of this mysterious indifference; but I apprehend that we have only to revert to the political state of Europe in the year 1813, and the problem will be solved.

Amongst the reasons for supposing that the Emperor must have felt ill disposed towards the English philosopher, the following story has been told; which, as an anecdote, is sufficiently amusing; and I can state upon the highest authority, that it is moreover perfectly true.

It is well known that Bonaparte, during his whole career, was in the habit of personal intercourse with the *savans* of Paris, and that he not unfrequently attended the sittings of the Institute. Upon being informed of the decomposition of the alkalies, he asked, with some impetuosity, how it happened that the discovery had not been made in France?—"We have never constructed a Voltaic battery of sufficient power," was the answer. "Then," exclaimed Bonaparte, "let one be instantly formed, without any regard to cost or labour."

The command of the Emperor was of course obeyed; and, on being informed that it was in full action, he repaired to the laboratory to witness its powers; on his alluding to the taste produced by the contact of two metals, with that rapidity which characterised all his motions, and before the attendants could interpose any precaution, he thrust the extreme wires of the battery under his tongue, and received a shock which nearly deprived him of sensation. After recovering from its effects, he quitted the laboratory without making any remark, and was never afterwards heard to refer to the subject.

It is only an act of justice to state that Davy, during his residence in the French capital, so far from truckling to French politics, never lost an opportunity of vindicating with temper the cause of his own country. At the Théâtre de la Porte Saint Martin, a melodrame was got up, with the avowed intention of exposing the English character to the execration of the audience. Lord Cornwallis was represented as the merciless assassin of the children of Tippoo Saib. Davy was highly incensed at the injustice of the representation, and abruptly quitted the theatre in a state of great indignation.

Whatever objections might have existed in his mind, as to his attending a levee of the Emperor, they did not operate in preventing his being presented to the Empress at Malmaison; but he could not be prevailed upon to appear upon that occasion, in any other than a morning dress; and it was not until after repeated entreaty, and the assurance that he would not be admitted into the *Salle de reception*, that he consented to exchange a pair of half-boots that laced in front, and came over the lower part of his pantaloons, for black silk stockings and shoes. His constant answer to the remonstrances of his friends was, "I shall go in the same dress to Malmaison as that in which I called upon the Prince Regent at Carlton House."

The introduction of Sir Humphry and Lady Davy to the Empress Josephine, took place at Malmaison on the 30th of November. The only English present were, the Earl of Beverley, a *détenue;* General Sir Edward Paget, a prisoner of war, taken in Spain, and Mr. Underwood; and it was the first levee at which any of our countrymen had been introduced, with the exception of Mr. Underwood, who had been frequently in the habit of paying his court to the Empress, and to whom, indeed, he was indebted for those indulgences which have been already mentioned.

The persons present having arranged themselves in a semicircle, the Empress entered the *Salle de reception*, and in her usual gracious manner addressed each individual. After this court ceremony, her Majesty retired, having previously signified to a select few, her desire that they should follow her into the private apartment.

In the Boudoir, the conversation became general,

and turned upon certain works of art; and upon Lady Davy expressing, in very florid terms, her admiration of some beautifully embellished cups of Porcelain, which were stationed on the mantel-piece, her Majesty, with that good-nature which ever distinguished her, immediately presented her with a specimen.

The Empress then proposed that Lady Davy should on this occasion visit her conservatories, upon which it is well known she had lavished large sums, and was ambitious to be thought to possess all that was rare and curious. Lady Davy having expressed some apprehensions as to the coldness of the day, and appearing to be but thinly clad, one of the Dames du Palais was commanded to provide cloaks; and in a short time, Mr. Underwood says, a *mountain* of the most costly and magnificent furs, that probably ever appeared even in a Regal Palace, were displayed before her; the splendid trophies, we may conclude, of the Royal conciliation at Tilsit.

It was on the 13th of December 1813, that Davy was elected a corresponding member of the First Class of the Imperial Institute; there were forty-eight members present, and he had forty-seven votes: Guyton de Morveau being the only person who opposed his election.

Nothing ever exceeded the liberality and unaffected kindness and attention with which the *savans* of France had received and caressed the English philosopher. Their conduct was the triumph of Science over national animosity,—a homage to genius, alike honourable to those who bestowed, and to him who received it; and it would be an act of ingratitude, a violation of historical justice, on the part of the English biographer, did he omit to express the pride and admiration with which every philosopher in his country continues to regard it. It would have been fortunate for the cause of Science, and fortunate for the historian, could he have terminated the subject with these remarks; but the biographer has an act of justice to perform, which he must not suffer his friendship to evade, nor his partialities to compromise.*

It would be an act of literary dishonesty to assert that Sir H. Davy returned the kindness of the *savans* of France, in a manner which the friends of Science could have expected and desired. There was a flippancy in his manner, a superciliousness and hauteur in his deportment, which surprised as much as they offended. Whatever opinions he might have formed as to the talents of the leading chemists, it was weakness to betray, and arrogance to avow them.

He had, by a single blow, fatally mutilated the system which was the pride and glory of their nation: it was ungenerous to remind them of his triumph. It required but little tact to have reconciled the French philosophers to the revolution

* In offering these observations, the reader may readily suppose it has not been without much pain that I have made this sacrifice of personal feeling to principle. I am, however, bound to observe, that Sir Humphry's sentiments towards France for the liberal indulgence granted to him were both grateful and kindly; and so strongly does Lady Davy participate in that feeling, that I perhaps owe it to her to state that neither her Ladyship's journals or information have been used upon this occasion.

he had effected; but, unfortunately, that cannot be said of Davy, which was so wittily observed of Voltaire,—that if he trod upon the toes of their prejudices, he touched his hat at the same time: even the affair of Iodine, had it been skilfully managed, would never have left an angry feeling. It was not his success, but the manner in which he spoke of it, that rendered it so offensive. He should have acted according to the judicious advice given to a member of the clerical profession, upon his consulting a friend as to the propriety of continuing his field-sports, should he become a dignitary of the Church—"You may hunt, but you must not holla."

It may be supposed that the unguarded conduct of Davy reached the ear of the Emperor; for in a conversation with one of the leading members of the Institute, Napoleon took occasion to observe, that he understood the young English chemist held them all in low estimation.

Having thus candidly avowed the errors of Davy, I may be justified in claiming from the reader his confidence in the sincerity with which I shall attempt to palliate them. From my personal knowledge of his character, I am inclined to refer much of that unfortunate manner, which has been considered as the expression of a haughty consciousness of superiority, to the desire of concealing a *mauvaise honte* and *gaucherie*—an ungraceful timidity, which he could never conquer. The bashful man, if he possess strong passions, will frequently force himself into a state of effrontery, by a violence of effort which passes amongst ordinary observers for the sallies of pride, or the ebullitions of temper; whereas if, on the contrary, his temperament be cold and passionless, he will exhibit traits of the most painful reserve. This proposition cannot, perhaps, be more forcibly illustrated than by a comparison of the manners of Davy and Cavendish, whose temperaments were certainly as much opposed to each other as fire is to ice: the latter, however, was shy and bashful, to a degree bordering upon disease; and nothing so much distressed him as an introduction to strangers, or as his being pointed out as a person distinguished in science. On one of the Sunday evening *soirées* of Sir Joseph Banks, he happened to be conversing with his friend Mr. Hatchett, when Dr. Ingenhouz, who was rather remarkable for pomposity of manner, approached him with an Austrian gentleman in his hand, and introduced him formally to Mr. Cavendish. He recounted the titles and qualifications of his foreign friend at great length, and concluded by saying, that he had been particularly anxious to be introduced to a philosopher so universally celebrated throughout Europe as Mr. Cavendish. As soon as Dr. Ingenhouz had finished, the Austrian gentleman began; he assured Mr. Cavendish that one of his principal inducements in coming to London, was to see and converse with one whom he considered the most distinguished chemist of the age. To all these high-flown addresses, Mr. Cavendish answered not a single word, but stood with his eyes cast down upon the floor, in a state of the most painful confusion. At length, espying an opening in the crowd, he darted through it with all the speed he could command, and never stopped until he reached his carriage, which immediately drove him home.

From the same cause, probably, arose Davy's inattention and carelessness in those little observances of etiquette, which many may treat as empty and unmeaning ceremonials, but which the members of a polished community regard as the delicate expressions of feeling, and the language of sentiment.

It is said, that on conversing in the chamber of the Institute, he received one of its most distinguished and venerable members, who approached him with the air of salutation, without rising from his seat; a circumstance perhaps in itself of very trifling importance, but it was considered as a mark of disrespect, which is not readily forgiven, when a spirit of rivalry may be supposed to sharpen the affront. It will be remembered that Cæsar might date his loss of popularity to the fact of his having received the Senate while sitting in the porch of the temple of Venus, and that it formed one of the chief pretences of those who organised the conspiracy against his person.

There were, besides, other sources of unpopularity, which we are bound in fairness and candour to impute to the excellencies, rather than to the defects, of his character. If we believe with Johnson, that men have sometimes gained reputation from their foibles, we may certainly admit the converse of the proposition, that they have occasionally lost it from their virtues. Davy, as we have seen, possessed from his earliest years a frankness of disposition which endeared him to all his friends, but in after life it unquestionably exposed him to various annoyances, which by a little reserve he would have certainly escaped. It is quite surprising how much a little mystery, judiciously managed, will achieve. Seven veils converted the fragment of a tile, ploughed up in the neighbourhood of Florence, into an object of awful devotion.*

Although it must be admitted that our philosopher lost some popularity during his visit to the French metropolis, the *savans* did not the less respect his talents, or admire his discoveries. They appear to have been impressed with the same sentiment as that which animated Voltaire, when he asked whether the discovery of Racine's weakness made the part of Phædra less admirable.

M. Dumas, who is certainly by no means distinguished for the readiness with which he is disposed to pay homage to British talent, has declared that Davy was the greatest chemical genius that ever appeared.

In fact, the more the researches of this great experimentalist are studied, the more they must be admired: every attempt to depreciate their intrinsic importance will only serve to display their exalted merits; every attempt to falsify their results will only tend to demonstrate their accuracy. It is by an elaborate examination only, that the full evidence of their truth can be displayed; there are points which the keen eye of genius will discern, that are invisible to a grosser sense: the coldness of criticism then will only make them glow the brighter; like his own potassium, the contact even of ice, so far from extinguishing, will light them up in splendour.

* Gray's Letters.

Sir Humphry left Paris on the morning of the 29th of December, and proceeded by the way of Lyons to Montpellier, where he remained for a month, and became acquainted with M. BERARD, who afterwards filled the chemical chair in that university, and in whose laboratory he worked upon the subject of Iodine, and examined many of the marine productions of the Mediterranean, with the view of determining whether they contained that body. M. Berard directed a considerable quantity of the species of *Ulva*, which abounds on the coast of Languedoc, to be burnt for him; and although the ashes consisted for the most part of common salt, he obtained traces of Iodine in the lixivium. From the general results of his experiments, however, he concludes, that the ashes of the *fuci* and *ulvæ* of the Mediterranean afford it in much smaller quantities than the sea-weed, from which soda is procured; and it was only in a very few instances that he could derive any evidence of its existence. In the ashes of the corallines and sponges, he could not obtain the slightest indication of its presence. During this period he also extended his enquiries respecting the chemical agencies of Iodine, and the properties of several of its compounds, especially of those in which he believed it to exist in triple combination with alkalies and oxygen, and for which he proposed the name of *Oxy-iodes*.*

While at Montpellier, Davy witnessed the pro-

* These are the *Iodates* of the present day; but Davy, it would seem, resisted the conviction of Iodic acid being an *oxy-acid*, upon the same grounds that he opposed the views of M. Gay-Lussac with regard to the nature of Chloric acid.

cession of the Pope on his return to Rome. His Holiness appeared in a state of great humiliation, and, on being supplicated by a poor woman to cure her child, he replied, that she must propitiate Heaven by her prayers, for that he was himself a mere mortal, without power to heal or to save.

He quitted Montpellier on the 7th of February, and, accompanied by M. Berard, visited the fountain of Vaucluse; he afterwards continued his route to Nice, crossed the Col de Tende, to Turin, and arrived at Genoa on the 25th of February; from which place the following letter is dated.

TO MR. UNDERWOOD.

MY DEAR UNDERWOOD, Genoa, March 4.

I HAVE not received the letter you announced to me in the street, concerning Ampère's note, nor any others since I left Paris. The note came to me through the Prefect of Nice, with an indorsement by M. Degerand.

I crossed the Alps by the Col de Tende, stayed at Turin three days, and came here through snow and ice, over the Bochetta, where I have been waiting for a fair wind for Tuscany. We have had no impediments except from the snow and the east winds.

If you can hear any thing of the destination of the letters I have twice missed, I shall thank you to let me know by addressing me at Rome at the *Posta*. I shall be most happy to hear some news of you here, and shall always feel a lively interest in your plans, and in your welfare.

I have been making some experiments here on

the Torpedo, but without any decisive results; the coldness of the weather renders the powers of the animal feeble; I hope, however, to resume them at Naples.

Tell M. Ampère, I hope he will not give up the subject of the laws of the combination of gaseous bodies, which is so worthy of being illustrated by his talents, and which offers such ample scope for his mathematical powers, united as they are with chemical knowledge:—tell him also that I hope he will sometimes write to me, and that I shall always remember with pleasure the hours I have passed in his society.

Pray tell me that you are well; and remember me to all that are interested in me. My wife desires her kind remembrances. I am, my dear Underwood,

Your very sincere friend,

H. Davy.

Besides researches on the Torpedo, Davy made farther experiments on the ashes of Sea-weed, which were collected for him by Professor Viviani, of Genoa.

He left Genoa by water on the 13th, and arrived at Florence on the 16th of March. Here he worked in the laboratory of the Academia del Cimento, on Iodine; but more particularly on the combustion of the Diamond. The experiments on this latter body were performed by means of the great lens in the cabinet of Natural History; the same instrument as that employed in the first trials on the action of the solar heat on the diamond, instituted by Cosmo III. Grand Duke of Tuscany: upon this occasion, he was assisted by COUNT BARDI, the Director, and SIGNIOR GAZZARI, the Professor of Chemistry at the Florentine Museum.

I have been informed that the hasty, and apparently careless manner in which he conducted his experiments, and which has been already noticed* as being characteristic of his style of manipulation, greatly astonished the philosophers of Florence, and even excited their alarm for the safety of the lens, which on all occasions had been used by them with such fastidious caution and delicacy.

In the very first trials on the combustion of the diamond, he ascertained a very curious circumstance that had not been before noticed; namely, that the diamond, when strongly ignited by the lens in a thin capsule of platinum, perforated with many orifices, so as to admit a free circulation of air, will continue to burn in oxygen gas after being withdrawn from the focus. The knowledge of this circumstance enabled him to adopt a very simple apparatus and mode of operation in his researches, and to complete in a few minutes experiments which had been supposed to require the presence of a bright sunshine for many hours.

The new facts obtained by the experiments on Iodine, which he had commenced at Montpellier and carried on at Florence, he embodied in a memoir, which was read before the Royal Society on the 16th of June 1814. It treated more particularly of the triple compounds containing iodine and oxygen,—of the hydrionic acid, and of the compounds procured by means of it,—of the combinations of

* Vol. i. p. 144.

iodine and chlorine,—of the action of some compound gases on iodine,*—and, lastly, of the mode of detecting iodine in combinations. "If iodine," he says, "exists in sea water, which there is every reason to believe must be the case, though in extremely minute quantities, it is probably in triple union with oxygen and sodium, and in this case it must separate with the first crystals of common salt."

He quitted Florence on the 3rd, and having visited Sienna, entered Rome on the 6th of April. The Continent having now become accessible, he met with many of his English friends: but neither the extended society by which he was surrounded, nor the classical attractions of the city of the Cæsars, allured him from the pursuits of Science. We find that, shortly after his arrival, he renewed his researches on the combustion of different kinds of charcoal, in the laboratory of the Academia del Lyncei, in which he was assisted by Sig. Morrichini and Barlocci, Professors of the College Sapienza at Rome. Having arranged the results of this investigation, together with those relating to the combustion of the diamond, which he had previously obtained at Florence, he transmitted a paper to the Royal Society, entitled "Some Experiments on the Combustion of the Diamond, and other carbonaceous substances;" which was read on the 23rd of June, and published in the Second Part of the Philosophical Transactions for the year 1814.

* The compounds which he supposed to be thus produced are of a very questionable nature; with respect to that formed with the Olefiant gas, he was evidently in error.

No sooner had it been established by various accurate experiments, that the diamond and common charcoal consumed nearly the same quantity of oxygen in combustion, and produced a gas having the same obvious qualities, than various conjectures were formed to explain the remarkable differences in the sensible qualities of these bodies, by supposing some minute difference in their chemical composition. MM. Biot and Arrago, from the high refractive power of the diamond, suspected that it might contain hydrogen. Guyton Morveau inferred from his experiments that it was pure carbon, and that charcoal was an oxide of carbon; whereas Davy was inclined to believe, from the circumstance of the non-conducting power of the diamond, as well as from the action of potassium upon it, that a minute portion of oxygen might enter its composition, although such a supposition would be at variance with the doctrine of definite proportions; but more lately, in his account of some new experiments on the fluoric compounds, he hazarded the idea that it might be the carbonaceous principle combined with some new and subtile element, belonging to the same class as oxygen, chlorine, and fluorine, which has hitherto escaped detection, but which may be expelled, or newly combined, during its combustion in oxygen. "That some chemical difference," says Davy, "must exist between the hardest and most beautiful of the gems and charcoal, between a non-conductor and a conductor of electricity, it is scarcely possible, notwithstanding the elaborate experiments that have been made on the subject, to doubt: and it seems reasonable to expect, that a very refined or perfect chemistry will confirm the analogies of Nature, and show that bodies cannot be exactly the same in composition or chemical nature, and yet totally different in all their physical properties."

With these impressions, we may readily imagine the ardour with which he availed himself of the use of the great lens at Florence. He had in various ways frequently attempted to fuse charcoal,* but without success. In a letter addressed to Mr. Children is the following passage: "The great result to be hoped for is the fusion of carbon; and then you may use diamond in the manufacture of gunpowder."

He tells us that he had long felt a desire to make some new experiments on the combustion of the diamond and other carbonaceous substances; and that this desire was increased by the new fact ascertained with respect to iodine, which by uniting to hydrogen, affords an acid so analogous to muriatic acid, that it was for some time confounded with that body. His object in these new experiments, was to examine minutely whether any peculiar matter was separated from the diamond during its combustion, and to determine whether the gas, formed in this process, was precisely the same in its minute chemical nature, as that formed in the combustion of common charcoal. By his experiments at Florence, he satisfactorily accomplished his wishes, and established beyond a question the important fact, that

* The supposed fusion of charcoal by Professor Silliman, by means of Dr. Hare's galvanic deflagrator, was a fallacy arising from the earthy impurities of the substance. See *American Journal of Science*, vol. v. p. 108, and 361.

"the diamond affords no other substance by its combustion than pure carbonic acid gas; and that the process is merely a solution of diamond in oxygen, without any change in the volume of the gas."

As one of the principal objects in these researches was to ascertain whether water was formed during the combustion of the diamond, with a view to decide the question of the presence of hydrogen, every possible source of fallacy was excluded. In one experiment there was an evident deposition of moisture, but it was immediately discovered to have been owing to the production of vapour from a cork connected with a part of the apparatus, during the combustion.

In the progress of this research, he ascertained a fact, the knowledge of which must not only be considered as important to the present enquiry, but as highly valuable in excluding error from our reasonings upon the delicate results of analysis*—I allude to the extremely minute quantity of water which becomes perceptible by deposition on a polished glass surface. He introduced a piece of paper weighing a grain into a tube of about the capacity of four cubical inches, the exterior of which was gently heated by a candle; immediately a slight but perceptible dew appeared in the interior of the upper part of the tube; the paper taken out and directly weighed in a balance, sensible to 1-100th of a grain, had not suffered any appreciable diminution. If,

* It has a more especial bearing upon that experimental research by which the nature of chlorine was established, as described at page 337, vol. i. to which I beg to refer the chemical reader.

then, on burning 1·84 grains of diamond in oxygen gas, not even a barely perceptible dew was produced, we may consider it as fully proved that this gem cannot contain hydrogen in its composition: but to render the demonstration, if possible, still more complete, he kept a small diamond, weighing ·45 of a grain, in a state of intense ignition by the great lens of the Florentine Museum, for more than half an hour, in chlorine; but the gas suffered no change, and the diamond underwent no alteration either in weight or appearance: now had the smallest portion of hydrogen been developed, white fumes of muriatic acid would have been visible, and a certain condensation of the gas must have taken place.

The general tenor of his results was equally opposed to the idea of the diamond containing oxygen; for, in such a case, the quantity of carbonic acid generated by the combustion, would, on comparison, have indicated that fact. By combining the carbonic acid with lime, and then recovering the gas from the precipitate by muriatic acid, he found its proportion to be exactly that which was furnished by an equal weight of Carrara marble similarly treated.

The enquiry next proceeds to the examination of other forms of carbonaceous matter, such as plumbago, charcoal formed by the action of sulphuric acid on oil of turpentine, and that produced during the formation of sulphuric ether; and lastly, the common charcoal of oak.

In all these bodies, he detected the presence of hydrogen, both by the water generated during their combustion, and by the production of muriatic acid, when ignited in chlorine. The chemical difference then between the diamond and the purest charcoal, would appear to consist in the latter containing hydrogen; but Davy very justly asks whether a quantity of an element, less in some cases than 1-5000th part of the weight of the substance, can occasion so great a difference in physical and chemical characters? "It is certainly possible," says he, "yet it is contrary to analogy, and I am more inclined to adopt the opinion of Mr. Tennant, that the difference depends upon crystallization." In support of such an opinion, he farther adduces the fact, that charcoal after being intensely ignited in chlorine, is not altered in its conducting power or colour: in which case the carbon is freed from the hydrogen, and yet undergoes no alteration in its physical properties.

One distinction supposed to exist between the diamond and common carbonaceous substance, the researches of Davy have certainly removed, viz. its relative inflammability; for he has shown that the former will burn in oxygen with as much facility as plumbago.

The experiments, then, which Davy conducted at Florence and Rome, have removed several important errors with regard to the nature of carbonaceous substances; and though they may not encourage the labours of those speculative chemists who still hope to illustrate the old proverb,* by manufacturing diamonds out of charcoal, they certainly show that they are less chimerical than those of the wild visionaries who sought to convert the baser metals into gold.

* "Carbonem pro Thesauro."

While at Rome, Davy was engaged for several successive days in the house of Morrichini, for the purpose of repeating with that philosopher his curious experiments on *magnetisation*. Mr. Faraday was charged with the performance of the experiments, but never could obtain any results.

On the 8th of May he entered Naples, and remained there for three weeks, during which period he visited Mount Vesuvius, and the volcanic country surrounding it. He describes the crater, at this time, as presenting the appearance of an immense funnel, closed at the bottom, with many small apertures emitting steam; while on the side towards Torre del Greco, there was a large aperture from which flame issued to a height of at least sixty yards, producing a most violent hissing noise. He was unable to approach sufficiently near the flame to ascertain the results of the combustion; but a considerable quantity of steam ascended from it; and he says, that when the wind blew the vapours upon him, there was a distinct smell both of sulphurous and muriatic acids, but there was no indication of carbonaceous matter from the colour of the smoke; nor was any deposited upon the yellow and white saline matter which surrounded the crater, and which he found to be principally sulphate and muriate of soda, and in some specimens there was also a considerable quantity of muriate of iron. At this period, when the volcano was comparatively tranquil, he observed the solfaterra to be in a very active state, throwing up large quantities of steam, and some sulphuretted hydrogen.

At several subsequent periods he revisited Vesuvius; and I shall hereafter take occasion to relate all the principal observations he made, and the conclusions at which he arrived, with respect to this the most interesting of all the phenomena of mineral nature.

He also took great interest in the excavations at that time going on at Pompeii, under the direction of Murat, then King of Naples, who placed at his disposal several specimens of art, which Davy received with a view to investigate the chemical composition of the colours used by the Ancients.

On the 25th of May he returned to Rome, and again quitted it on the 2nd of June.

I regret to say that the information I have received, as to the future continental travels of our philosopher, is extremely meagre, and will consist of little more than names and dates. Of this, however, the reader may be assured, that nothing which relates to his scientific researches has been omitted.

From Rome he proceeded to Terni, and thence to Bologna, where he remained for three days; then to Mantua, Verona, and Milan. Whether at this or at some subsequent period he went to Pavia, in order to pay his homage to the illustrious Volta, I entertain some doubt; but the time is immaterial to the point of the anecdote I am about to relate.

Davy had sent a letter to Pavia to announce his intended visit; and on the appointed day and hour, Volta, in full dress, anxiously awaited his arrival. On the entrance of the great English philosopher into the apartment, not only in *déshabille*, but in a dress of which an English artisan would have been ashamed, Volta started back in astonishment, and such was the effect of his surprise, that he was for some time unable to address him.

From Milan, which he left on the 22nd of June, he went to Como, Domo D'Ossola, and then over the Simplon, to Geneva, where he arrived on the 25th of that month, and remained until the 18th of September. During this visit he made various experiments on Iodine, at the house of De Saussure, which was situated near the edge of the Lake, and about three miles from Geneva. He also worked at M. Pictet's house, on the subject of the heat in the solar spectrum. Here also he met with a number of celebrated persons, whose society he greatly enjoyed; amongst whom were Madame de Stael, Benjamin Constant, Necker, and Talma. Lausanne, Vevay, Payerne, Berne, Zurich, Schaffhausen, and Munich, were successively visited by him. His route was then continued through Tyrol, Inspruck, Calmar, Bolsenna, Trent, Bassano, Vicenza, Padua, to Venice; where having remained two days, he returned to Padua on the 16th of October, and then proceeded to Ferrara, Bologna, and Pietra-Mala; near which latter place, in the Apennines, he examined a fire produced by gaseous matter constantly disengaged from a schist stratum, and from the results of its combustion, he concluded it to be pure fire-damp. On again reaching Florence, he found that the Professors had been dismissed, but he nevertheless resumed his researches, first at home, and afterwards in the laboratory of the Grand Duke, where he submitted to analysis some gas which had been collected by his attendant Mr. Faraday, from a cavity in the earth, about a mile from

Pietra-Mala, then filled with water, and which from the quantity of gas disengaged is called *Aqua Buja.* It was found to be pure light carburetted hydrogen, requiring two volumes of oxygen for its combustion, and producing a volume of carbonic acid gas. "It is very probable," says he, "that these gases were disengaged from coal strata beneath the surface, or from bituminous schist above coal; and at some future period new sources of wealth may be opened to Tuscany from this invaluable mineral treasure."

On the 29th he left Florence, and passing through Levano, Tortona, and Terni, arrived again at Rome, on the 2nd of November, where he remained till the 1st of March 1815.

During this winter he was engaged in an elaborate enquiry into the composition of ancient colours; and also in experiments upon certain compounds of iodine and of chlorine. Upon which subjects he transmitted to the Royal Society three memoirs, viz. one entitled, "Some Experiments and Observations on the Colours used in Painting by the Ancients," which was read on the 23rd of February; a second, "On a solid compound of Iodine and Oxygen," read April 10; and a third, "On the action of Acids on the Salts usually called Hyper-oxymuriates, and on the Gases produced from them," read May 4, 1815; all of which were published in the Philosophical Transactions for that year.

Although the paintings of the great masters of Greece have been entirely destroyed, either by accident, by time, or by the barbarian conquerors at the period of the decline and fall of the Roman

Empire, yet there is sufficient proof that this art attained a very high degree of excellence amongst a people to whom genius and taste were a kind of birthright, and who possessed a perception, which seemed almost instinctive, of the dignified, the beautiful, and the sublime.

Our philosopher observes, that the subjects of many of those pictures are described in ancient authors, and that some idea of the manner and style of the Greek artists may be gained from the designs on the vases improperly called "Etruscan," which were executed by artists of Magna Græcia, and many of which are probably copies from celebrated works: of their execution and colouring, some faint notion may be gained from the paintings in fresco found at Rome, Herculaneum, and Pompeii; for, although these paintings are not properly Grecian, yet at the period when Rome was the metropolis of the world, the fine arts were cultivated in that city exclusively by Greek artists, or by artists of the Greek school; while it is evident, on comparing the descriptions of Vitruvius and Pliny with those of Theophrastus, that the same materials for colouring were employed at Rome and at Athens.

With regard to the nature of these pigments, we may obtain some information from the works of Theophrastus, Dioscorides, Vitruvius, and Pliny; but until the present memoir by Sir H. Davy, no experimental attempt had been made to identify them, or to imitate such of them as are peculiar.

His experiments, he informs us, were made upon colours found in the Baths of Titus, and the ruins called the Baths of Livia, and in the remains of other palaces and baths of ancient Rome, and in the ruins of Pompeii.

By the kindness of his friend Canova, who was charged with the care of the works connected with ancient art in Rome, he was enabled to select, with his own hands, specimens of the different pigments that were found in vases discovered in the excavations made beneath the ruins of the palace of Titus, and to compare them with the colours fixed on the walls, or detached in fragments of stucco; and Signor Nelli, the proprietor of the "Nozze Aldobrandine,"* permitted him to make such experiments upon the colours of that celebrated picture, as was necessary to determine their nature.

Without entering into the chemical details of the subject, I shall offer a general history of the nature of the colours he examined.

Of the red colours, he distinguished four distinct kinds, viz.—one bright and approaching to orange, which he found to be *Minium*, or the red oxide of lead; a second, dull red, which he ascertained to be an iron ochre; a third, a purplish red, which was likewise an ochre, but of a different tint; and a fourth, a brighter red than the first, which was *Vermilion* or *Cinnabar*, a sulphuret of mercury. On examining the fresco paintings in the Baths of Titus, he found that all the three first colours had been used, the ochres particularly, in the shades of the figures, and the minium in the ornaments on

* The most celebrated picture of antiquity rescued from the ruins of Herculaneum. It represents a virgin on her marriage night, with her female attendants. An engraving of it is to be seen in Sir William Hamilton's work on Herculaneum.

the borders. The fourth red had been employed in various apartments, and formed the basis of the colouring of the niche, and of other parts of the chamber in which the Laocoon is said to have been found in the time of Raphael; a circumstance which Davy considers as being favourable to the belief that such apartments were intended for Imperial use, since vermilion, amongst the Romans, was a colour held in the highest esteem, and was always one of great costliness.

Of the yellows, the more inferior were mixtures of ochre and different quantities of chalk; the richer varieties were ochres mixed with the red oxide of lead.

The ancients had also two other colours, which were orange, or yellow; the auripigmentum, or αρσενικον, said to approach to gold in the brilliancy of its tint, and which is described by Vitruvius as being found native in Pontus, and which Davy says was evidently sulphuret of arsenic;—and a pale sandarach, said by Pliny to have been found in gold and silver mines, and which was imitated at Rome by a partial calcination of cerusse. He conceives that this must have been *Massicot,* or the yellow oxide of lead mixed with minium; I suspect, however, that Davy was mistaken in supposing that the ancients always applied the term Sandarach to minium; the Σανδαρακη of Aristotle was evidently an arsenical sulphuret.

In his examination of the ancient Frescoes, he could not detect the use of orpiment; but a deep yellow, approaching to orange, which covered a piece of stucco in the ruins near the monument of

Caius Cestius, proved to be oxide of lead, and consisted of massicot and minium. He considers it probable that the ancients used many colours from lead of different tints, between the "*usta*" of Pliny, which was our minium, and imperfectly decomposed cerusse, or pale massicot.

The differently shaded blues, by the action of acids, uniformly assumed the same tint; from which he concluded that the effect of the base was varied by different proportions of chalk. This base he ascertained to be a *frit*, made by means of soda and sand, and coloured by oxide of copper.

The greens were, in general, combinations of copper; and it seemed probable, that although they appeared in the state of carbonate, they might originally have been laid on in that of acetate. The purple of the ancients, the πορφορα of the Greeks, and the *Ostrum* of the Romans, was regarded as their most beautiful colour, and was obtained from shell-fish. Vitruvius states that it was prepared by beating the fish with instruments of iron, freeing the purple liquor from the shell containing it, and then mixing it with a little honey. Pliny says that, for the use of the painters, *argentine creta*, (probably a clay used for polishing silver,) was dyed with it, and both Vitruvius and Pliny state that it was adulterated, or imitations of it made by tinging *creta* with madder; whence it would appear, that the ancients were acquainted with the art of making a lake colour from that plant, similar to the one used by modern painters.

Pliny informs us, that the finest purple had a tint like a deep-coloured rose. In the Baths of Titus, there was found a broken vase of earthenware, which contained a pale rose colour; and Davy selected it as an appropriate subject for his analytical experiments.

Where this colour had been exposed to the action of the air, its tint had faded into a cream colour, but the interior parts retained a lustre approaching to that of carmine. A diluted acid was found to dissolve out of it a considerable quantity of carbonate of lime, with which the colouring principle must have been mixed, as a substance of a bright rose colour remained after the process. This colouring ingredient was proved to contain siliceous, aluminous, and calcareous earths, without any sensible trace of metallic matter, except oxide of iron. Upon heating the substance, first in oxygen, and then with hyperoxymuriate of potash, Davy was induced to consider the colouring matter itself as either of vegetable or animal origin; the results, however, were so equivocal, that he renounced the hope of determining its nature from the products of its decomposition. If it be of animal origin, he thinks it is most probably the Tyrian or marine purple, as it is likely that the most expensive colour would have been employed in ornamenting the Imperial baths.

He had not observed any colour of the same tint as this ancient lake in the fresco paintings; the purplish reds in the Baths of Titus he ascertained to be mixtures of red ochres and the blues of copper.

The blacks and browns were mixtures of carbonaceous matter, with the ores of iron or manganese. The black from the Baths of Titus, as well as that from some ruins near the Porta del Popolo, deflagrated with nitre, and presented all the character of carbon. This fact agrees with the statements of all the ancient authors who have described the artificial Greek and Roman black as consisting of carbonaceous matter, either prepared from the powder of charcoal, from the decomposition of resin, (a species of lamp black,) from that of the lees of wine, or from the common soot of wood fires. Pliny also mentions the inks of the cuttle-fish, but adds, "*Ex his non fit.*"

Davy informs us, that, some years before, he had examined the black matter of the cuttle-fish, and had found it to be a carbonaceous substance mixed with gelatine.*

Pliny, moreover, speaks of ivory black invented by Apelles; of a natural fossil black; and of a black prepared from an earth of the colour of sulphur. Davy is of opinion, that both these latter pigments were ores of iron and manganese; and he observes that the analysis of some purple glass satisfied him that the ancients were well acquainted with the ores of manganese.

The *whites* which he examined from the Baths of Titus, as well as those from other ruins, were either chalk, or fine aluminous clay; and he states that, amongst all his researches, he never once met with cerusse.

This interesting account of the colours used by the ancients is followed by observations on the manner in which they were applied; and the paper is concluded with some general remarks of much practical importance.

* I find from a note addressed by Davy to Mr. Underwood, that he was engaged in these experiments in October 1801.

The azure, he says, of which the excellence is sufficiently proved by its duration for 1700 years, may be easily and cheaply imitated: he found, for instance, that fifteen parts of carbonate of soda, twenty parts of opaque flint powdered, and three parts of copper filings, by weight, when strongly heated together for two hours, yielded a compound substance of exactly the same tint, and of nearly the same degree of fusibility; and which, when powdered, produced a fine deep sky-blue.

The azure, the red and yellow ochres, and the blacks, appear to have been the only pigments which have not undergone any change in the fresco paintings. The vermilion presents a darker hue than that of recently made Dutch cinnabar; and the red lead is inferior in tint to that sold in the shops. The greens are generally dull.

The blue frit above mentioned, he considers as a colour composed upon the truest principles; and he thinks there is reason to believe, that it is the colour described by Theophrastus as the one manufactured at Alexandria. "It embodies," says he, "the colour in a composition like stone, so as to prevent the escape of elastic matter from it, or the decomposing action of the elements upon it." He suggests the possibility of making other *frits*, and thinks it would be worth while to try whether the beautiful purple given by oxide of gold could not be made useful in a deeply tinted glass.

Where *frit* cannot be employed, he observes that metallic combinations which are insoluble in water, and which are saturated with oxygen or some acid matter, have been proved by the testimony of seventeen centuries to be the best pigments. In the red ochres, for example, the oxide of iron is fully combined with oxygen and carbonic acid; and the colours composed of them have never changed. The carbonates of copper, which consist of an oxide and an acid, have suffered but little alteration. Massicot and orpiment, he considers as those which have been the least permanent amongst all the mineral colours.

He next takes a view of the colours which owe their origin to the improvements of modern chemistry. He considers the *patent yellow* to be more permanent, and the chromate of lead more beautiful, than any yellow possessed by the Greeks or Romans. He pronounces *Scheele's green* (arsenite of copper), and the insoluble muriatic combinations of copper, to be more unalterable than the ancient greens; and he thinks that the sulphate of baryta offers a white far superior to any pigment possessed by the ancients.

In examining the colours used in the celebrated Nozze Aldobrandine, he recognised all the compounds which his analytical enquiries had established: viz. the reds and yellows were all ochres; the blues, the Alexandrian frit; the greens, copper; the purple, especially that in the garment of the Pronuba, appeared to be a compound colour of red ochre and copper; the browns and blacks were mixtures of ochres and carbon; while the whites were carbonate of lime.

" The great Greek painters," he adds, " like the most illustrious artists of the Roman and Venetian school, were probably sparing in the use of the more florid tints in historical and moral painting, and produced their effects rather by the contrasts of colouring in those parts of the picture where a deep and uniform tint might be used, than by brilliant drapery.

"If red and yellow ochres, blacks and whites, were the pigments most employed by Protogenes and Apelles, so they are likewise the colours most employed by Raphael and Titian in their best style. The St. John and the Venus, in the tribune of the Gallery at Florence, offer striking examples of pictures in which all the deeper tints are evidently produced by red and yellow ochres, and carbonaceous substances.

"As far as colours are concerned, these works are prepared for that immortality which they deserve; but unfortunately, the oil and the canvass are vegetable materials, and liable to decomposition, and the last is even less durable than the wood on which the Greek artists painted their celebrated pictures.

"It is unfortunate that the materials for receiving those works which are worthy of passing down to posterity as eternal monuments of genius, taste, and industry, are not imperishable marble or stone:* and that *frit*, or unalterable metallic combinations have not been the only pigments employed by great artists; and that their varnishes have not been sought for amongst the transparent compounds† unalterable in the atmosphere.

* Copper, it is evident from the specimens in the ruins of Pompeii, is a very perishable material; but modern science might suggest some voltaic protection.

† Davy thinks that the artificial hydrat of alumina will probably

In his memoir "On a solid compound of Iodine and Oxygen," he enumerates, amongst the agencies of that body, its singular property of forming crystalline combinations with all the fluid or solid acids. It will be unnecessary to follow him through this investigation, since its results have been found to be erroneous. M. Serullas* has lately shown that the crystalline bodies of Davy are nothing more than the iodic acid, which being insoluble in acids, is necessarily precipitated by them.

His paper "On the action of Acids on Salts usually called Hyper-oxymuriates," announced the important fact of chlorine forming with oxygen a compound, in which the latter element exists in a still greater proportion than in the body previously described by him under the name of *Euchlorine*.†

Before finally quitting Italy, he spent three weeks at Naples, during which period he experimented on iodine and fluorine in the house of Sementini; he also paid several visits to Vesuvius, and found the appearances of the crater to be entirely different from those which it presented in the preceding year:‡ there was, for instance, no aperture in it; it was often quiet for minutes together, and then burst out into explosions with considerable violence, sending fluid lava, and ignited stones and ashes, to a height of many hundred feet in the air.

"These eruptions," says he, "were preceded by

be found to be a substance of this kind; and that, possibly, the solution of boracic acid in alcohol will form a varnish. He also thinks, that the solution of sulphur in alcohol is worthy of an experiment.

* Annales de Chimie, t. 43. p. 216.

† See page 330. vol. i.

‡ See page 42. vol. ii.

subterraneous thunder, which appeared to come from a great distance, and which sometimes lasted for a minute. During the four times that I was upon the crater, in the month of March, I had at last learnt to estimate the violence of the eruption from the nature of the sound: loud and long-continued subterraneous thunder indicated a considerable explosion. Before the eruption, the crater appeared perfectly tranquil; and the bottom, apparently without an aperture, was covered with ashes. Soon, indistinct rumbling sounds were heard, as if at a great distance; gradually, the sound approached nearer, and was like the noise of artillery fired under our feet. The ashes then began to rise and to be thrown out with smoke from the bottom of the crater; and lastly, the lava and ignited matter was ejected with a most violent explosion. I need not say, that when I was standing on the edge of the crater, witnessing this phenomenon, the wind was blowing strongly from me; without this circumstance, it would have been dangerous to have remained in such a situation; and whenever from the loudness of the thunder the eruption promised to be violent, I always ran as far as possible from the seat of danger.

"As soon as the eruption had taken place, the ashes and stones which rolled down the crater seemed to fill up the aperture, so that it appeared as if the ignited and elastic matter were discharged laterally; and the interior of the crater assumed the same appearance as before."

On the 21st of March, he quitted Naples, and returned to England by the following route: Rome—Narni—Nocere—Fessombone—Imola—Mantua—(March 30,) Verona—Pero—Trente—Botzen—Brenńah—Inspruck—Zirl—(April 4,) Reuti-Menningen—Ulm—(April 6,) Stutgard—Heidelburg—Mayence—Boppert—Coblentz—Cologne—(April 14,) Leuch—Brussels—Ostend—Dover—London, April 23, 1815.

CHAPTER XI.

Collieries of the North of England.—Fire-damp.—The dreadful explosion at Felling Colliery described.—Letters from the Bishop of Bristol to the Author.—A Society is established at Bishop-Wearmouth for preventing accidents in coal mines.—Various projects for ensuring the miner's safety.—The Reverend Dr. Gray, the present Bishop of Bristol, addresses a letter to Sir H. Davy, and invites his attention to the subject.—Sir H. Davy's reply.—Farther correspondence upon the possibility of devising means of security.—Sir H. Davy proposes four different kinds of lamp for the purpose.—The Safe-lamp—The Blowing-lamp—The Piston-lamp—The Charcoal-lamp.—His investigation of the properties of fire-damp leads to the discovery of a new principle of safety.—His views developed in a paper read before the Royal Society on the 9th of November 1815.—The first Safety-lamp.—Safety-tubes superseded by Safety-canals.—Flame Sieves.—Wire-gauze lamp.—The phenomenon of slow Combustion, and its curious application.—The invention of the Safety-lamp claimed by a Mr. Stephenson.—A deputation of Coal-owners wait upon Sir H. Davy, in order to express to him the thanks of the Proprietors for his discovery.—Mr. Buddle announces to Dr. Gray (now Bishop of Bristol) the intention of the Coal trade to present him with a service of plate.—The Resolutions are opposed, and the claims of Stephenson urged, by Mr. W. Brandling.—A dinner is given to Sir Humphry, at which the plate is presented to him.—The President and Council of the Royal Society protest against the claims still urged by Mr. Stephenson's friends.—Mr. Buddle's letter in answer to several queries submitted to him by the Author.—Davy's Researches on Flame.—He receives from the Royal Society the Rumford Medals.—Is created a Baronet.—Some observations on the apathy of the State in rewarding scientific merit.—The Geological Society of Cornwall receives the patronage and support of Sir Humphry.

A FEW months after the return of Sir Humphry Davy to England, his talents were put in requisition to discover some remedy for an evil which had hitherto defied the skill of the best practical engineers and mechanics of the kingdom, and which continued to scatter misery and death amidst an important and laborious class of our countrymen.

To collect and publish a detailed account of the numerous and awful accidents which have occurred within the last few years, from the explosion of inflammable air, or *fire-damp*, in the coal mines of the North of England, would present a picture of the most appalling nature. It appears from a statement by Dr. Clanny, in the year 1813,* that, in the space of seven years, upwards of three hundred pitmen had been suddenly deprived of their lives, besides a considerable number who had been severely wounded; and that more than three hundred women and children had been left in a state of the greatest distress and poverty; since which period the mines have increased in depth, and until the happy discovery of Davy, the accidents continued to increase in number.

It may well be asked how it can possibly have happened that, in a country so enlightened by science and so distinguished for humanity, an evil of such fearful magnitude, and of such frequent recurrence, should for so long a period have excited but little sympathy, beyond the immediate scene of

* Phil. Transact. 1813.

the catastrophe. It would seem that a certain degree of doubt and mystery, or novelty, is essentially necessary to create that species of dramatic interest by which the passions are excited through the medium of the imagination: it is thus that the philanthropist penetrates unknown regions, in search of objects for his compassion, while he passes unheeded the miserable groups who crowd his threshold; it is thus that the statesman pleads the injuries of the Negro with an eloquence that shakes the thrones of kings, while he bestows not a thought upon the intrepid labourers in his own country, who for a miserable pittance pass their days in the caverns of the earth, to procure for him the means of defying the severity of winter, and of chasing away the gloom of his climate by an artificial sunshine.

That the benefits conferred upon mankind by the labours of Sir H. Davy may be properly appreciated, it is necessary to describe the magnitude of the evil which his genius has removed, as well as the numerous difficulties which opposed his efforts and counteracted his designs.

The great coal field,* the scene of those awful accidents which will be hereafter described, extends over a considerable part of the counties of Northumberland and Durham. The whole surface has been calculated at a hundred and eighty square

* Dr. Thomson has calculated that the quantity of coal exported yearly from this formation exceeds two millions of chaldrons; and he thinks it may be fairly stated, in round numbers, that, at the present rate of waste, it will continue to supply coal for a thousand years! Mr. Phillips, however, is inclined to deduct a century or two from this calculation.

miles, and the number of different beds of coal has been stated to exceed forty; many of which, however, are insignificant in point of dimensions. The two most important are about six feet in thickness, and are distinguished by the names of *High main*, and *Low main*, the former being about sixty fathoms above the latter.

From this statement, some idea may be formed of the great extent of the excavations, and of the consequent difficulty of successfully ventilating the mines. In some collieries, they are continued for many miles, forming numerous windings and turnings, along which the pitmen have frequently to walk for forty or fifty minutes, before they arrive at the workings; during which time, as well as when at work, they have no direct communication with the surface of the earth. The most ingenious machinery, however, has been contrived for conducting pure air through every part of the mine, and for even ventilating the old excavations, which are technically called *Wastes;* and unless some obstruction occur, the plan* so far answers, as to furnish wholesome air to the pitmen, and to diminish, although, for reasons to be hereafter stated, it can never wholly prevent, the dangers of *fire-damp;* the nature of which it will be necessary to consider.

* In all large collieries, the air is accelerated through the workings by placing furnaces, sometimes at the bottom, and sometimes at the top of the up-cast shaft; in aid of which, at Wall's-end Colliery, a powerful air-pump worked by a steam-engine is employed to quicken the draft: this alone draws out of the mines a thousand hogsheads of air every minute. Stoppings and trapdoors are also interposed in various parts of the workings, in order to give a direction to the draft.

The coal appears to part with a portion of *carburetted hydrogen,* when newly exposed to the atmosphere; a fact which explains the well-known circumstance of the coal being more inflammable when fresh from the pit, than after long exposure to the air. We are informed by the Rev. Mr. Hodgson, that, on pounding some common Newcastle coal fresh from the mine, in a cask furnished with a small aperture, he found the gas which issued from it to be inflammable; and Davy, on breaking some lumps of coal under water, also ascertained that they gave off inflammable gas. The supposition that the coal strata have been formed under a pressure greater than that of the atmosphere, may furnish a clue to the comprehension of this phenomenon.

On some occasions the pitmen have opened with their picks crevices, or fissures, in the coal or shale, which have emitted as much as seven hundred hogsheads of *fire-damp* in a minute. These *Blowers*, as they are technically termed, have been known to continue in a state of activity for many months, or even years together;* a phenomenon which clearly shows that the carburetted hydrogen must have existed in the cavities of the strata in a very highly compressed, if not actually in a liquid state, and which, on the diminution of pressure, has resumed its elastic form.

All the sources of carburetted hydrogen would appear to unite in the deep and valuable collieries situated between the great North road and the sea.

* Sir James Lowther found a uniform current of this description produced in one of his mines for the space of two years and nine months. Phil. Trans. vol. 38. p. 112.

Their air courses are thirty or forty miles in length; and here, as might be expected, the most tremendous explosions have happened. Old workings, likewise, upon being broken into, have not unfrequently been found filled with this gas, and which, by mingling itself with the common air, has converted the whole atmosphere of the mine into a magazine of *fire-damp.*

On the approach of a candle, it is in an instant kindled: the expanding fluid drives before it a roaring whirlwind of flaming air, which tears up every thing in its progress, scorching some of the miners to a cinder, and burying others under enormous ruins shaken from the roof; when thundering to the shafts, it converts the mine, as it were, into an enormous piece of artillery, and wastes its fury in a discharge of thick clouds of coal-dust, stones, and timber, together with the limbs and mangled bodies of men and horses.

But this first, though apparently the most appalling, is not the most destructive effect of these subterraneous combustions. All the *stoppings* and trapdoors of the mine being blown down by the violence of the concussion, and the atmospheric current entirely excluded from the workings, such of the miners as may have survived the discharge are doomed to the more painful and lingering death of suffocation from the *after-damp,* or *stythe,* as it is termed, which immediately results from the combustion, and occupies the vacuum necessarily produced by it.

As the phenomena accompanying these explosions are always of the same description, to relate the numerous recorded histories of such accidents would be only to multiply pictures of death and human suffering, without an adequate object: it is, however, essential to the just comprehension of the subject, that the reader should receive at least one well-authenticated account, in all its terrific details; and I have accordingly selected that which was originally drawn up with much accuracy and feeling by the Reverend John Hodgson, and which is prefixed to the funeral sermon preached on the occasion, and subsequently published by that gentleman.

The accident occurred at Felling Colliery, near Sunderland, on the 25th of May, in the year 1812. This mine was considered by the workmen as a model of perfection, both with regard to the purity of its air, and the arrangements of its machinery. The concern was in the highest degree prosperous; and no accident, except a trifling explosion which slightly scorched two or three pitmen, had ever occurred.

Two *shifts*, or sets of men, were constantly employed, the first of which entered the mine at four, and were relieved at their working posts by the next set at eleven o'clock in the morning; but such was the confidence of the pitmen in the safety of this mine, that the second shift of men were often at their posts before the first set had left them; and such happened to be the case on the following unhappy occasion.

About half past eleven, on the morning of the 25th of May, the neighbouring villages were alarmed by a tremendous explosion. The subterraneous fire broke forth with two heavy discharges from the shaft called the '*John Pit*,' which was one hundred and two fathoms deep, and were almost immediately followed by one from that termed the '*William pit*.' A slight trembling, as if from an earthquake, was felt for about half a mile around the workings; and the noise of the explosion, though dull, was heard to the distance of three or four miles, and greatly resembled an unsteady fire of infantry.

Immense quantities of dust and small coal accompanied these blasts, and rose high into the air, in the form of an inverted cone. The heaviest part of the ejected matter, such as masses of timber, and fragments of coal, fell near the pit, but the dust, borne away by a strong west wind, fell in a continued shower to the distance of a mile and a half; and in the village of Heworth, it caused a gloom, like that of early twilight, and so covered the roads that the footsteps of passengers were strongly imprinted on them.

As soon as the explosion had been heard, the wives and children of the pitmen rushed to the working pit. Wildness and terror were pictured in every countenance. The crowd thickened from every side, and in a very short period several hundred persons had collected together; and the air resounded with exclamations of despair for the fate of husbands, parents, and children.

The machinery having been rendered useless by the eruption, the rope of the *gin* was sent down the shaft with all possible expedition. In the absence of horses, a number of men, who seemed to acquire strength as the necessity for it increased, applied their shoulders to the *starts*, or shafts of the gin, and worked it with astonishing expedition.

By twelve o'clock, thirty-two persons, all that survived this dreadful catastrophe, had been brought to day-light, but of these three boys lived only a few hours. The dead bodies of two boys, miserably scorched and shattered, were also brought up at the same time. Twenty-nine persons, then, were all who were left to relate what they had observed of the appearances and effect of the explosion. One hundred and twenty-one were in the mine when it happened, eighty-seven of whom remained in the workings. Eight persons had fortunately come up a short time before the accident.

Those who had their friends restored, hastened with them from the scene of destruction, and for a while appeared to suffer as much from an excess of joy, as they had a short time before from the depth of despair; while those who were yet in the agony of suspense, filled the air with shrieks and howlings, and ran about wringing their hands and throwing their bodies into the most frantic and extravagant gestures.

As not one of the pitmen had escaped from the mine by the only avenue open to them, the apprehension for their safety momentarily increased, and at a quarter after twelve o'clock, nine persons descended the John pit, with the faint hope that some might still survive.

As the fire-damp would have been instantly ignited by candles, they lighted their way by *steel-mills*;* and knowing that a great number of the

* Steel-mills are small machines, which give light by turning a cylinder of steel against a piece of flint. Sir James Lowther had observed early in the last century, that the fire-damp in its usual

miners must have been at the crane when the explosion happened, they at once attempted to reach that spot: their progress, however, was very soon intercepted by the prevalence of *choak damp*, and the sparks from their steel-mill fell into it like dark drops of blood: deprived therefore of light, and nearly suffocated by the noxious atmosphere, they retraced their steps towards the shaft, but they were shortly stopped by a thick smoke which stood like a wall before them. Here their steel-mills became entirely useless, and the chance of their ever finding any of their companions alive entirely hopeless; to which should also be added the horror arising from the conviction of the mine being on fire, and the probability of a second explosion occurring at the next moment, and of their being buried in the ruins it would occasion.

At two o'clock, five of the intrepid persons who had thus nobly volunteered their assistance, ascended; two were still in the shaft, and the other two remained below, when a second explosion, much less severe, however, than the first, excited amongst the relatives of those entombed in the mine still more frightful expressions of grief and despair. The persons in the shaft experienced but little inconvenience from this fresh eruption, while those below, on hearing the distant growlings, immediately threw themselves flat on their faces, and in this posture, by keeping a firm hold on a strong wooden prop, they felt no other annoyance from

form was not inflammable by sparks from flint and steel; and it appears that a person in his employment invented the machine in question.

the blast than that of having their bodies tossed in various directions, in the manner that a buoy is heaved by the waves of the sea. As soon as the atmospheric current returned down the shaft, they were safely drawn to the light.

As each came up, he was surrounded by a group of anxious enquirers; but not a ray of hope could be elicited; and the second explosion so strongly corroborated their account of the impure state of the mine, that their assertions for the present seemed to obtain credit. This impression, however, was but of short duration,—hope still lingered; they recollected that persons had survived similar accidents, and that, upon opening the mine, they had been found alive after considerable intervals. Three miners, for instance, had been shut up for forty days in a pit near Byker, and during the whole of that period had subsisted on candles and horse-beans. Persons too were not wanting to agitate the minds of the relatives with disbelief in the report of the pitmen who had lately descended to explore the mine. It was suggested to them, that want of courage, or bribery, might have induced them to magnify the danger, and to represent the impossibility of reaching the bodies of the unfortunate sufferers. By this species of wicked industry, the grief of the neighbourhood began to change its gloomy, for an irritable aspect. The proposition to exclude the atmospheric air from the mine, in order to extinguish the fire, was received with the cries of *Murder!*—and with the determination to oppose such a proceeding by violence.

Many of the widows lingered about the mouth of

the pit during the whole of the night with the hope of hearing the cries of a husband or a son.

On Tuesday the 26th, that natural propensity in the human mind to derive gratification from spectacles of horror, was exemplified in a very striking manner. An immense crowd of colliers from various parts, but more especially from the banks of the river Wear, assembled around the pit, and were clamorous in their reproaches of the persons concerned in the management of the mine, accusing them of want of perseverance in their attempts to rescue the unhappy sufferers. Every one had some successful adventure to relate; all were liberal in their professions of readiness to give assistance; but not one was found hardy enough to enter the jaws of the burning cavern.

The leaders of this outcry, however, who had been led into error by an impulse which did honour to their hearts, were soon brought to listen with patience to a relation of all the circumstances of the explosion, and of the reasons for concluding that the mine was then actually on fire, and the persons enclosed in it beyond the hope of recovery. They very candidly allowed, after this explanation, the impracticability of any attempt to reach the bodies of the sufferers, until the fire was extinguished; and they accordingly urged the propriety of excluding from the mine the access of air, as the only means of accomplishing the object. At the same time, the proprietors gave the strongest assurances to the multitude, that if any project could be devised for the recovery of their friends, no cost or labour should be spared in executing it; that, if any person could be found willing to enter the mine, every facility and assistance should be afforded him; but, as they were assured by the most eminent Viewers that the workings were inaccessible, they would not hold out any reward for the attempt,—they would not be accessary to the death of any one, either by persuasion or bribery.

At the clamorous solicitation, however, of the populace, two persons again descended the shaft, and very nearly lost their lives in the attempt. The report of these last adventurers, in a great measure, convinced the people of the impossibility of their friends' survival in so deadly an atmosphere, and reconciled them to the plan of excluding the air. The operation was accordingly commenced, and it proceeded without interruption; but from various accidents, more than a month elapsed before the mine was in a state to admit of examination; and during this interval, numerous were the idle tales which had been circulated throughout the country. Several of the sufferers, it was said, had found their way to the shafts, and been recovered. Their number even had been circumstantially told—how they had subsisted on candles, oats, and beans, and how they had heard the different persons who explored the mine in the hope of rescuing them.

Some conjuror too, it was said, had set his spells and divinations to work, and had penetrated all the secrets of the mine. He had discovered one famishing group receiving drops of water from the roof, another eating their shoes and clothes, and many other similar pictures of horror. These inventions were carefully related to the widows, and they produced the effect of daily harrowing up afresh their sorrows; indeed, it seemed the chief employment of some to indulge in a kind of insane sport with their own and their neighbours' calamity.

The morning of Wednesday the 8th of July having been appointed for exploring the workings, the distress of the neighbourhood was again renewed at an early hour: a great concourse of people assembled; some, out of curiosity, to witness the commencement of an undertaking full of sadness and danger,—some to excite the revenge, or to aggravate the sorrows, of the relatives by calumnies and reproaches, for the sole purpose of mischief; but the greater part came with broken hearts and streaming eyes, in expectation of seeing the mangled corpse of a father, brother, husband, or son.

The *shifts* of men employed in this doleful and unwholesome work were generally about eight in number. They were four hours in, and eight hours out of the mine; so that each individual wrought two shifts every twenty-four hours.

When the first shift of men came up, a message was dispatched for a number of coffins to be in readiness at the mouth of the pit. Ninety-two had been prepared, and they had to pass by the village of Low Felling, in their way to the mine. As soon as a cart-load of them was seen, the howling of the women, who, hitherto secluded in their dwellings, had now begun to assemble about their doors, came on the breeze in slow fitful gusts, which presaged a scene of the greatest distress and confusion.

The bodies were found under various circumstances: one miner, from his position, must have been sleeping when the explosion happened, and had never opened his eyes. In one spot were found twenty-one bodies in ghastly confusion,—some like mummies, scorched as dry as if they had been baked; one wanted its head, another an arm—the scene was most terrific: the power of the fire was visible upon all, but its effects were very various; while some were almost torn to pieces, there were others who appeared as if they had sunk down overpowered by sleep.

Every family had made arrangements for receiving the dead bodies of their kindred; but Dr. Ramsay having given his opinion, that such a proceeding might spread a putrid fever through the neighbourhood, and the first body when exposed to observation having presented a most horrid and corrupt appearance, the people very properly consented to have each body interred as soon as it was discovered, on condition that the hearse, in its way to the chapel-yard, should pass by the door of the deceased.

From the 8th of July to the 19th of September, the heart-rending scene of mothers and widows examining the putrid bodies of their sons and husbands, for marks by which to identify them, was daily renewed; but very few of them were recognised by any personal mark—they were too much mangled and scorched to retain any of their features: their clothes, tobacco-boxes, and shoes, were therefore the only indications by which they could be identified.

The total loss from this terrible accident was

ninety-two pitmen; while forty widows, sixty girls, and twenty-six boys, comprising in all one hundred and twenty six persons, were thrown upon the benevolence of the public.

It was impossible that an event of such awful magnitude should not have deeply affected every humane person resident in the district. Nothing, in short, could exceed the anxiety which was manifested on the occasion; but, most unfortunately, there existed an invincible prejudice against every proposition that could be offered, from a general impression as to the utter hopelessness of any attempt to discover a remedy. A few philosophic individuals, however, did form themselves, as we shall presently learn, into an association for the laudable purpose of inviting the attention of scientific men to the subject, and of obtaining from them any suggestions which might lead to a more secure method of lighting the mines.

To the Reverend Dr. Gray, the present Lord Bishop of Bristol, who, at the period to which I allude, was the Rector of Bishop-Wearmouth, and one of the most zealous and intelligent members of the association, I beg to offer my public acknowledgments and thanks for the several highly interesting communications and letters with which his Lordship has obliged me, and by means of which I have been enabled to present to the scientific world a complete history of those proceedings which have so happily led to a discovery, of which it is not too much to say that it is, at once, the pride of science, the triumph of humanity, and the glory of the age in which we live.

In a letter I had lately the honour of receiving from that learned prelate, his Lordship says, "It was at a time when all relief was deemed hopeless, that Mr. Wilkinson, a barrister in London, and a gentleman distinguished for the humanity of his disposition, suggested the expediency of establishing a society for the purpose of enquiring whether any, and what, methods of security might be adopted for the prevention of those accidents so frequently occurring in the collieries of Northumberland and Durham.

"In consequence of this benevolent suggestion, a society was established at Bishop-Wearmouth, on the 1st of October 1813, by Sir Ralph Milbanke, afterwards Sir Ralph Noel, Dr. Gray, Dr. Pemberton, Mr. Robinson, Mr. Stephenson, and several other gentlemen. It was entitled, 'A Society for preventing Accidents in Coal-Mines;' and it immediately obtained the patronage of the Bishop of Durham, the Duke of Northumberland, and other noblemen and gentlemen.

"A very few days before the first meeting, twenty-seven persons had been killed in a colliery in which Sir Ralph Milbanke had an interest, and he was called upon at the meeting to state the particulars of the accident. At that time there was such little expectation that any means could be devised to prevent the occurrence of these explosions, that the object of the gentlemen who convened the meeting, however humane in principle, was considered by the persons present as chimerical and visionary. The Society, however, amidst many difficulties and considerable discouragement, and a perpetual harass by the offer of impracticable schemes from every quarter, nevertheless persevered in their meetings, and succeeded in establishing a communication and correspondence with other Societies in different parts of the kingdom."*

* It is unnecessary to enumerate the various schemes that have been proposed to prevent accidents from fire-damp. Some were unquestionably of value, and might, by their adoption, have diminished the frequency of explosions; others were visionary, or wholly impracticable. It was proposed, for instance, to fill the mine with an atmosphere of chlorine, which by entering into chemical union with the carburetted hydrogen, might disarm it of its power. Dr. Murray, in a paper published in the Transactions of the Royal Society of Edinburgh, suggests the use of a lamp that shall be supplied with air from the ground of the pit, by means of a long flexible tube, upon the false assumption that the fire-damp alone occupies the higher parts of the mine. Mr. W. Brandling also constructed a Safe-lamp, which, like that of Dr. Murray, was fed by air introduced through a long flexible tube reaching to the floor of the mine. In addition to which, he attached to the top of the lantern a pair of double bellows, by the aid of which he at the same time drew out the contaminated air from the interior of the lamp, and sucked in, through the flexible tube, a fresh portion to supply its place. To say nothing of the inefficacy and inconvenience of the long tube, the bellows possessed the additional objection of frequently puffing out the light.

One of the most active and intelligent members of the "Society for preventing accidents in Coal Mines," Dr. Clanny, had for some time paid particular attention to the object in contemplation. He first suggested the idea of an insulated lamp, of which an account appeared in the Philosophical Transactions for 1813. In 1815, he invented a steam safety-lamp, constructed of the strongest tinned iron, with thick flint glass in front. In this machine, the air of the coal mine passes in a current through a tube, and mixing with the steam, before it can arrive at the light, burns steadily in the wick of the lamp alone. This lamp has the valuable property of remaining cool. It has been much used in the Herrington Mill pit, the Whitefield pit, and the Engine pit.

"One of the projects offered was, that electrical machines should be employed, with ramifications to extend through all the departments of the collieries, and which were to be excited in discharging their fluid in constant succession, in order gradually to destroy the inflammable air. Many other suggestions were proposed, the principal of which were formed with the intention of purifying the air of the pits by chemical processes, or by forcing in large quantities of atmospheric air, through pipes and tunnels, &c.

"The Society, although it received some distinguished patronage, was not furnished with means sufficiently ample for exciting emulation by premiums, or even for defraying the expenses of intelligent artisans; and it unfortunately lost a considerable portion of its funds by the failure of the Wear Bank.

"Amongst the applications which more particularly excited the attention of the Society, was that of Mr. Ryan of Donegal, who objected to the principle upon which the working of collieries was carried on. He conceived that they should be originally constructed at the commencement of the working, with a view to admit the escape of the hydrogen gas to the highest parts of the colliery. He proposed to ventilate even the foulest pits, and the attention of the gentlemen proprietors, or occupiers of collieries, in the neighbourhood of Newcastle, was called upon at public meetings, and an enquiry set on foot with respect to the validity of his pretensions. Some gentlemen were even deputed to proceed into Staffordshire to ascertain the nature and extent of his services in that county, where he had been for some time employed. An offer was also made to place under his management the Hecton pit, at Hepburn, which was particularly foul; but a difference of opinion having arisen as to the efficacy of his plan, he did not consider himself sufficiently encouraged to proceed, and he left the country dissatisfied. He afterwards received the gold medal from the Society for promoting the Arts and Sciences."

The Society having as yet effected but little towards the great object of their deliberations, the chairman of the committee, Dr. Gray, who was generally acquainted with Sir Humphry Davy, judged it expedient to direct his attention to a subject, upon which, of all men of science, he appeared to be the best calculated to bring his extensive stores of chemical knowledge to a practical bearing.

As the life of this valuable man is now closed, and as every incident in it is interesting as connected with the advancement of philosophical knowledge, and especially of chemical discoveries important to the welfare of mankind, it may be worth while to enter into a review of the proceedings which were adopted upon this occasion, in order to trace the progress of the discoveries which were made, and the methods by which he arrived at his conclusions.

Dr. Gray, the chairman of the committee, having addressed to him a letter with a view to engage him in an investigation so important to society, received from him the following answer.

TO THE REVEREND DR. GRAY.

SIR, August 3, 1815.

I HAD the honour of receiving the letter which you addressed to me in London, at this place, and I am much obliged to you for calling my attention to so important a subject.

It will give me great satisfaction if my chemical knowledge can be of any use in an enquiry so interesting to humanity, and I beg you will assure the Committee of my readiness to co-operate with them in any experiments or investigations on the subject.

If you think my visiting the mines can be of any use, I will cheerfully do so.

There appears to me to be several modes of destroying the fire-damp without danger; but the difficulty is to ascertain when it is present, without introducing lights which may inflame it. I have thought of two species of lights which have no power of inflaming the gas which is the cause of the fire-damp, but I have not here the means of ascertaining whether they will be sufficiently luminous to enable the workmen to carry on their business. They can be easily procured, and at a cheaper rate than candles.

I do not recollect any thing of Mr. Ryan's plan: it is possible that it has been mentioned to me in general conversation, and that I have forgotten it. If it has been communicated to me in any other way, it has made no impression on my memory.

I shall be here for ten days longer, and on my return South, will visit any place you will be kind enough to point out to me, where I may be able to acquire information on the subject of the coal gas.

Should the Bishop of Durham be at Auckland, I shall pay my respects to his Lordship on my return.

I have the honour to be, dear Sir, with much respect, your obedient humble servant,

H. DAVY.

At Lord Somerville's, near Melrose, N. B.

TO THE SAME.

SIR, Melrose, August 18, 1815.

I RECEIVED your letter, which followed me to the Moors, where I have been shooting with Lord Somerville. I should have replied to it before this time, but we were in a part of the Highlands where there was no post. I am very grateful to you for the obliging invitation it contains.

I propose to leave the Tweed side on Tuesday or Wednesday, so that I shall be at Newcastle either on Wednesday or Thursday. If you will have the kindness to inform me by a letter, addressed at the Post Office, where I can find the gentleman you mention, I will call upon him, and do any thing in my power to assist the investigation in that neighbourhood.

I regret that I cannot say positively whether I shall be at Newcastle on Wednesday or Thursday; for I have some business at Kelso which may detain me for a night, or it may be finished immediately.

I am travelling as a bachelor, and will do myself the honour of paying my respects to you at Bishop-Wearmouth towards the end of the week.

I am, Sir, with much respect,

Your obedient humble servant,

H. DAVY.

The gentleman alluded to in the preceding letter, and to whom Dr. Gray wished Sir H. Davy to apply, was Mr. Buddle, a person whose extensive practical knowledge has justly entitled him to be considered as the highest authority on all subjects connected with the art of mining, and who has conferred inestimable benefits on the mining interests by the introduction of successful methods of ventilation. The account of his interview with Sir H. Davy is communicated in the following letter.

MR. BUDDLE TO DR. GRAY.

Wall's-end Colliery, August 24, 1815.

SIR,

PERMIT me to offer my best acknowledgments for the opportunity which your attention to the cause of humanity has afforded me of being introduced to Sir Humphry Davy.

I was this morning favoured with a call from him, and he was accompanied by the Rev. Mr. Hodgson. He made particular enquiries into the nature of the danger arising from the discharge of the inflammable gas in our mines. I shall supply him with a quantity of the gas to analyze; and he has given me reason to expect that a substitute may be found for the steel mill, which will not fire the gas. He seems also to think it possible to generate a gas, at a moderate expense, which, by mixing with the atmospheric current, will so far neutralize the inflammable air, as to prevent it firing at the candles of the workmen.

If he should be so fortunate as to succeed in either the one or the other of these points, he will render the most essential benefit to the mining interest of this country, and to the cause of humanity in particular.

I have little doubt but it will be gratifying to you to be informed, that progress is making towards the establishment of a permanent fund for the relief of sufferers by accident and sickness in the collieries of this district.

I have the pleasure to remain, with great respect, Sir, your most obedient, humble servant,

JOHN BUDDLE.

Sir H. Davy on his return to London, having been supplied by Mr. Buddle with various specimens of *fire-damp*, proceeded, in the first instance, to submit to a minute chemical examination the substance with which he had to contend.

In less than a fortnight, he informed Dr. Gray by letter, that he had discovered some new and unexpected properties in the gas, which had led to no less than four different plans for lighting the mines with safety.

TO THE REVEREND DR. GRAY.

MY DEAR SIR, Royal Institution, Oct. 30.

As it was the consequence of your invitation that I endeavoured to investigate the nature of the fire-damp, I owe to you the first notice of the progress of my experiments.

My results have been successful far beyond my expectations. I shall enclose a little sketch of my views on the subject; and I hope in a few days to be able to send a paper with the apparatus for the committee.

I trust the *Safe lamp* will answer all the objects of the collier.

I consider this at present as a *private* communication. I wish you to examine the lamps I have had constructed, before you give any account of my labours to the committee.

I have never received so much pleasure from the result of any of my chemical labours; for I trust the cause of humanity will gain something by it.

I beg of you to present my best respects to Mrs. Gray, and to remember me to your son.

I am, my dear Sir, with many thanks for your hospitality and kindness when I was at Sunderland, your obliged servant,

H. DAVY.

TO THE SAME.

MY DEAR SIR, London, October 31, 1815.

I SENT yesterday a sketch of my results on the fire-damp. We have lately heard so much of East* Shields, that by a strange accident I confounded it with Bishop-Wearmouth, and addressed your letter to East Shields.

I could not find any body to frank it, and you will find it a heavy packet; however, I could not lose a moment in giving you an account of results which I hope may be useful to humanity.

If my letter has not reached you, it will be found at the Post Office, East Shields.

With respects to Mrs. Gray, I am, my dear Sir, very sincerely yours,

H. DAVY.

* *Quere*—South Shields.

The sketch alluded to in the foregoing letter, has been kindly placed in my hands by the Bishop of Bristol; it possesses considerable interest as an original document, displaying his earliest views, and tending to illustrate the history of their progress.

"The fire-damp I find, by chemical analysis, to be (as it has been always supposed) a hydro-carbonate. It is a chemical combination of hydrogen gas and carbon, in the proportion of 4 by weight of hydrogen gas, and 11½ of charcoal.

"I find it will not explode, if mixed with less than six times, or more than fourteen times its volume of atmospheric air. Air, when rendered impure by the combustion of a candle, but in which the candle will still burn, will not explode the gas from the mines; and when a lamp or candle is made to burn in a close vessel having apertures only above and below, an *explosive mixture* of gas admitted *merely enlarges* the light, and then gradually extinguishes it without explosion. Again,—the gas mixed in any proportion with common air, I have discovered, *will not explode* in a *small tube*, the diameter of which is less than ⅛th of an inch, or even a larger tube, if there is a mechanical force urging the gas through this tube.

"Explosive mixtures of this gas with air require much stronger heat for their explosion than mixtures of common inflammable gas.* Red-hot char-

* *Olefiant* gas, when mixed with such proportions of common air as to render it explosive, is fired both by charcoal and iron heated to a dull-red heat. *Gaseous oxide of carbon*, which explodes when mixed with two parts of air, is likewise inflammable by red-

coal, made so as not to flame, if blown up by a mixture of the mine gas and common air, does not explode it, but gives light in it; and iron, to cause the explosion of mixtures of this gas with air, must be made *white*-hot.

"The discovery of these curious and unexpected properties of the gas, leads to several practical methods of lighting the mines without any danger of explosion.

"The first and simplest is what I shall call the *Safe lamp*, in which a candle or a lamp burns in a safe lantern which is air-tight in the sides, which has tubes below for admitting air, a chamber above, and a chimney for the foul air to pass through; and this is as portable as a common lantern, and not much more expensive. In this, the light never burns in its full quantity of air, and therefore is more feeble than that of the common candle.

"The second is the *Blowing lamp*. In this, the candle or lamp burns in a close lantern, having a tube below of small diameter for admitting air, which is thrown in by a small pair of bellows, and a tube above of the same diameter, furnished with a cup filled with oil. This burns brighter than the simple safe lamp, and is extinguished by explosive mixtures of the fire-damp. In this apparatus the candle may be made to burn as bright as in the air; and supposing an explosion to be made in it, it cannot reach to the external air.

"The third is the *Piston lamp*, in which the can-

hot iron, and charcoal. The case is the same with *sulphuretted hydrogen*.

dle is made to burn in a small glass lantern furnished with a piston, so constructed as to admit of air being supplied and thrown into it without any communication between the burner and the external air: this apparatus is not larger than the steel-mill, but it is more expensive than the other, costing from twenty-two to twenty-four shillings.

"These lamps are all extinguished when the air becomes so polluted with fire-damp as to be explosive.

"There is a fourth lamp, by means of which any *blowers* may be examined in air in which respiration cannot be carried on: that is, the *Charcoal lamp*. This consists of a small iron cage on a stand, containing small pieces of *very well burnt* charcoal blown up to a red heat. This light will not inflame any mixtures of air with fire-damp.*

"Of these inventions, the *Safe lamp*, which is the simplest, is likewise the one which affords the most perfect security, and requires no more care or attention than the common candle, and when the air in mines becomes improper for respiration, it is

* "In addition to these four lamps, we learn from an Appendix to his Paper in the Philosophical Transactions, that in the beginning of his enquiries, he constructed a close lantern, which he called the *Fire-valve lantern*; in which the candle or lamp burnt with its full quantity of air, admitted from an aperture below, till the air began to be mixed with fire-damp, when, as the fire-damp increased the flame, a thermometrical spring at the top of the lantern, made of brass and steel, riveted together, and in a curved form, expanded, moved a valve in the chimney, diminished the circulation of air, and extinguished the flame. He did not, however, pursue this invention, after he had discovered the properties of the fire-damp, on which his Safety-lamp is founded.

extinguished, and the workmen ought immediately to leave the place till a proper quantity of atmospheric air can be supplied by ventilation.

"I have made many experiments on these lamps with the genuine fire-damp taken from a blower in the Hepburn Colliery, collected under the inspection of Mr. Dunn, and sent to me by the Reverend Mr. Hodgson. My results have been always unequivocal.

"I shall immediately send models of the different lamps to such of the mines as are exposed to danger from explosion; and it will be the highest gratification to me to have assisted by my efforts a cause so interesting to humanity."

Contrary to the wish expressed by Sir Humphry Davy, the foregoing communication was inadvertently read at a public meeting of the Coal-trade, which was held at Newcastle on the 3rd of November: a circumstance which occasioned some embarrassment at the time, but is satisfactorily explained in the following letter from Sir Humphry.

TO THE REVEREND DR. GRAY.

MY DEAR SIR, 23, Grosvenor Street, Dec. 14, 1815.

MY communication to —— was, like that I made to you, intended to be *private;* he has however written to me to apologize for having made it known at Newcastle, stating, that having seen a notice of my results in the paper, the motive, as he conceived, for withholding it was at an end, as he considered my only reason for wishing to keep back my results from the public eye was the conviction that they might be rendered more perfect, and this I have now fully proved.

I trust I shall be able in a very few days to send you a model of a lantern nearly as simple as a common glass lantern, and which *cannot* communicate explosion to the fire-damp. I will send another to Newcastle, and I will likewise send you the copy of my paper, which you may reprint in any form you please; you will find my acknowledgments to you publicly stated.

My principles are these: *First*, a certain mixture of azote and carbonic acid prevents the explosion of the fire-damp, and this mixture is necessarily formed in the safe lantern; — *Secondly*, the fire-damp *will not explode* in tubes or feeders of a certain small diameter. The ingress into, and egress of air from my lantern is through such tubes or feeders; and therefore, when an explosion is *artificially* made in the safe lantern, it does not communicate to the external air.

I have made two or three lanterns of different forms. Experience must determine which will be the most convenient.

Should there be a little delay in sending them, it will be the fault of the manufacturer. It is impossible to conceive the difficulty of getting any thing made in London which is not in the common routine of business; and I should be very sorry to send you any thing imperfectly executed.

With best respects to Mrs. Gray, I am, my dear Sir,

Very sincerely your obliged servant,

H. DAVY.

The paper alluded to in the preceding letter, entitled, "On the Fire-damp of Coal Mines, and on methods of lighting the mine so as to prevent its explosion," was read before the Royal Society on the 9th of November, 1815.

In this memoir he communicates the results of some chemical experiments upon the nature of the fire-damp, and announces the existence of certain properties in that gas, which had previously escaped observation, and which leads to very simple methods of lighting the mines without danger.

He confirms the opinion of Dr. Henry, and other chemists, as to the fire-damp being light carburetted hydrogen gas, and analogous to the inflammable gas of marshes; but he found that the degree of its combustibility differed most materially from that of the other common inflammable gases, which it is well known will explode by the contact of both red-hot iron and charcoal; whereas well-burned charcoal, ignited to the strongest red heat, did not explode any mixture of the air and of the fire-damp; and a fire made of well-burned charcoal, that is to say, of charcoal that will burn without flame, was actually blown up to whiteness by an explosive mixture containing the fire-damp without producing its inflammation.* An iron rod also, at the highest degree of *red* heat, and even at the common degree of *white* heat, did not inflame explosive

* Whence he observes that, if it be necessary to be present in a part of the mine where the fire-damp is explosive, for the purpose of clearing the workings, taking away pillars of coal, or other objects, the workmen may be safely lighted by a fire made of charcoal, which burns without flame.

mixtures of the fire-damp; but when in brilliant combustion, it produced the effect.

He moreover found that the heat produced by the combustion of the fire-damp was much less than that occasioned by most other inflammable gases under similar circumstances; and hence its explosion was accompanied with comparatively less expansion: a circumstance of obvious importance in connection with the propagation of its flame.

Numerous experiments were likewise instituted by him with a view to determine the proportions of air with which the fire-damp required to be mixed, in order to produce an explosive atmosphere; and he found the quantity necessary for that purpose to be very considerable; even when mixed with three or nearly four times its bulk of air, it burnt quietly in the atmosphere, and extinguished a taper. When mixed with between five and six times its volume of air, it exploded freely. The mixture which seemed to possess the greatest explosive power was that of seven or eight parts of air to one of gas.

On adding azote and carbonic acid in different proportions to explosive mixtures of fire-damp, it was observed that, even in very small quantities, these gases diminished the velocity of the inflammation, or altogether destroyed it. In this stage of the enquiry, the important fact was discovered, that explosive mixtures could not be fired in metallic tubes of certain lengths and diameters.* In ex-

* Mr. Tennant had, some years before, observed that mixtures of the gas, from the distillation of coal, and air, would not explode in very small tubes. Davy, however, was not aware of this at the time of his researches.

ploding, for instance, a mixture of one part of gas from the distillation of coal, and eight parts of air, in a tube of a quarter of an inch in diameter and a foot long, more than a second was required before the flame reached from one end of the tube to the other; and not any mixture could be made to explode in a glass tube of one-seventh of an inch in diameter. In pursuing these experiments, he found that, by diminishing its diameter, he might in the same ratio shorten the tube without danger; and that the same principle of security was obtained by diminishing the length and increasing the number of the tubes, so that a great number of small apertures would not pass explosion when their depth was equal to their diameter. This fact led him to trials upon sieves made of wire-gauze, or metallic plates perforated with numerous small holes, and he found that it was impossible to pass explosions through them.*

In reasoning upon these several phenomena, it occurred to him, that as a high temperature was required for the inflammation of the fire-damp, and as it produced in burning, comparatively, *a small degree* of heat, the effect of carbonic acid and azote, as well as that of the surfaces of small tubes, in preventing its explosion, depended upon their cooling powers; that is to say, upon their lowering the temperature of the exploding mixture to such a degree,

* The apertures in the gauze should not be more than one-twentieth of an inch square. As the fire-damp is not inflamed by ignited wire, the thickness of the wire is not of importance; but wire from one-fortieth to one-sixtieth of an inch in diameter is the most convenient.

that it was no longer sufficient for its continuous inflammation. In support of this theory, he ascertained that metallic tubes resisted the passage of the flame more powerfully than glass tubes of similar lengths and diameters, metal being the better conductor of heat; and that carbonic acid was more effective than azote in depriving the fire-damp of its explosive power, in consequence, as he considered, of its greater capacity for heat, and likewise of a higher conducting power connected with its greater density.

In this short statement, the reader is presented with the whole theory and operation of the Safety-lamp, which is nothing more than an apparatus by which the inflammable air, upon exploding in its interior, cannot pass out without being so far cooled, as to deprive it of the power of communicating inflammation to the surrounding atmosphere. The principle having been once discovered, it was easy to adopt and multiply practical applications of it.

From the result of these researches, it became at once evident, that to light mines infested with fire-damp, with perfect security, it was only necessary to use an air-tight lantern, supplied with air from tubes of small diameter, through which explosions cannot pass, and with a chimney, on a similar principle, at the upper part, to carry off the foul air. A common lantern, to be adapted to the purpose, merely required to be made air-tight in the door and sides, and to be furnished with the chimney, and the system of safety apertures below and above the flame of the lamp. Such, in fact, was Davy's first Safety-lamp; and having afterwards varied the arrangement of the tubes in different ways, he at length exchanged them for canals, which consisted of close concentric hollow metallic cylinders of different diameter, so placed together as to form circular canals of the diameter of from one-twenty-fifth to one fortieth of an inch; and of an inch and seven-tenths in length; by which air is admitted in much larger quantities than by the small tubes, and they are moreover much superior to the latter in practical application. He also found, that longitudinal air-canals of metal might be employed with the same security as circular canals; and that a few pieces of tin plate, soldered together with wires to regulate the diameter of the canal, answered the purpose of the feeder or safe chimney, as well as drawn cylinders of brass.

The subjoined explanatory sketches will assist in rendering the scheme intelligible, and obviate the possibility of any misconception of the subject.

Fig. 1. represents the first Safe lantern, with its air-feeder and chimney furnished with safety metallic canals. The sides are of horn or glass, made air-tight by putty or cement. A. is the lamp through which the circular air-feeding canals pass. B. is the chimney containing four such canals; above it is a hollow cylinder, with a cap to prevent dust from passing into the chimney. C. is the hole for admitting oil. F. is the rim round the bottom of the lantern, to enable it to bear motion.

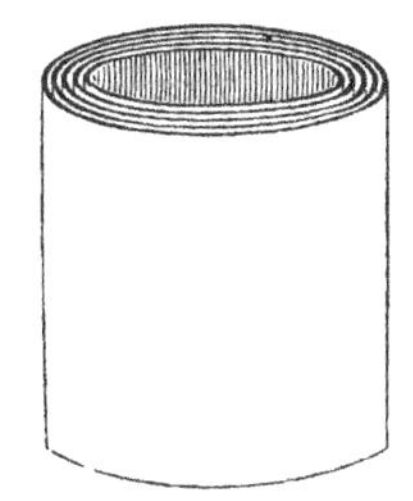

FIG. 2. exhibits an enlarged view of the safety concentric canals, which, if one-twenty-fifth of an inch in diameter, must not be less than two inches in exterior circumference, and one-seventh of an inch high.

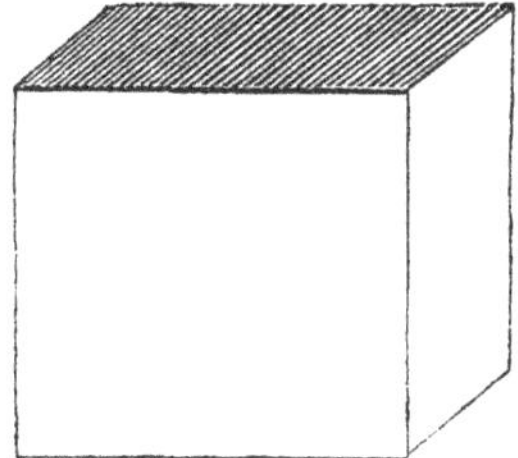

FIG. 3. exhibits the longitudinal safety canals.

FIG. 4. represents a Safety-lamp having a glass chimney, covered with tin-plate, and the safety apertures in a cylinder with a covering above: the lower part is the same as in the lantern.

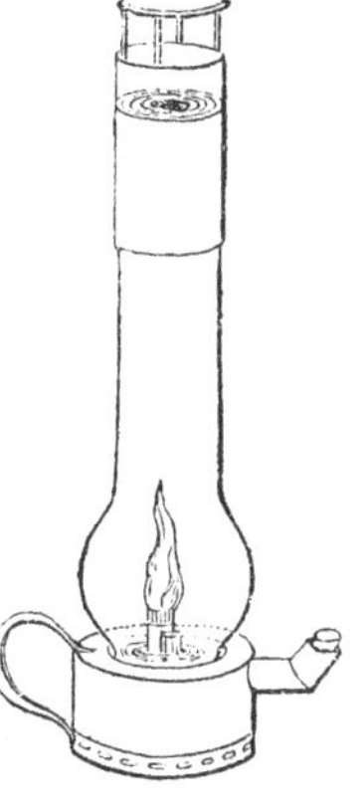

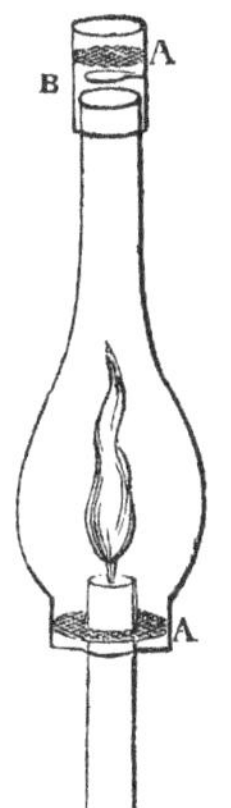

FIG. 5. A glass tube furnished with *flame sieves*, in which a common candle may be burnt. A A. the flame sieves. B. a little plate of metal to prevent the upper flame sieve from being acted on by the current of hot air.

During the short visit of Sir Humphry Davy at Bishop Wearmouth, he saw the lamp which Dr. Clanny was then engaged in perfecting. It has been already observed, that it was secured against the effects of fire-damp by being supplied with atmospheric air previously conveyed through water.* The machinery of this lamp was far too cumbrous to be of general use; but its inventor was justly commended by Davy for his ingenuity and perseverance. It unfortunately happened that, in consequence of some erroneous representations made to Dr. Clanny, he received the impression that Sir Humphry had not been disposed to treat his invention with sufficient respect, nor had given him the credit to which he was so justly entitled. This suspicion, which had been long industriously kept alive, was however ultimately removed.

The following letter refers to this unfortunate circumstance. I have adverted to it in these memoirs, for the purpose of showing what an unfair spirit of rivalry, and what a succession of petty jealousies were excited by those generous and disinterested labours of Davy, which ought to have called forth nothing but the most lively expressions of gratitude for his services, and admiration of his genius.

TO THE REVEREND DR. GRAY.

MY DEAR SIR, 23, Grosvenor Street, December 13.

A FRIEND of mine has sent me a newspaper—

* M. de Humboldt conceived and executed the plan of a lamp in 1796, for giving a safe light in mines, upon a similar principle of entire insulation from the air.—*Journal des Mines*, t. viii. p. 839.

the Tyne Mercury, containing a very foolish libel upon me. It states, amongst other things, that I did not mention Dr. Clanny, or his lamp, in my late paper read before the Royal Society; whereas I mentioned his lamp as a very ingenious contrivance, and named him amongst the gentlemen who obligingly furnished me with information upon the subject.

It will be needless for me to point out to you that my lamp has no one principle in common with that of Dr. Clanny. He forces in his air through water by bellows. In mine, the air passes through safety canals without any mechanical assistance. Mine is a common lantern made close, and furnished with safety canals.

I hope I shall not hear that Dr. Clanny has in any way authorized or promoted so improper a statement as that in the Tyne Mercury; indeed, I do not think it possible.

I have at last obtained a complete model of my lamp, after many disappointments from the instrument-maker. I hope in a few days to send you a *Safe lantern*, as portable as a common-made one, and the perfect security of which is demonstrable.

I am, my dear Sir, your sincerely obliged,

H. DAVY.

TO THE SAME.

MY DEAR SIR, Grosvenor Street, December 15.

I SHALL inclose the first sheet of my paper, and shall be glad to preface it by some observations when you reprint it.

I shall forward my lanterns and lamps to you in

a few days. They are *absolutely* safe; and if the miners have any more explosions from their light, it will be their own fault.

You will find, when you see my construction, that the principles as well as the execution are entirely new.

You will find in the second sheet of my paper, which I hope to be able to send to-morrow, the *principles* of *security*, and its limits unfolded.

I am, my dear Sir, very sincerely yours,

H. DAVY.

TO THE SAME.

MY DEAR SIR, London, January 1, 1816.

I FEAR you will have accused me of procrastination in delaying to send you my papers and my lamps.

The papers read to the Royal Society have been printed; but during the period that has elapsed since I last wrote to you, I have made a discovery much more important than those which I have already had the honour of communicating to you.

I have made very simple and economical lanterns, and candle guards, which are not only *absolutely safe*, but which give light by means of the fire-damp, and which, while they disarm this destructive agent, make it useful to the miner.

This discovery is a consequence of that which I communicated to you in my last letter on the wire sieves. I hope to be able to send you on Wednesday the printed account of my results, together with models of lamps which will burn and consume all explosive mixtures of the fire-damp.

I have at last finished my enquiries with perfect satisfaction to myself, and I feel highly obliged to you for having called my attention to a subject where my labours will, I hope, be of some use.

I am, my dear Sir, very sincerely yours,

H. Davy.

It is impossible to approach the consideration of this last, the most signal and splendid of his triumphs, without feelings of the highest satisfaction. He had already, as we have seen, disarmed the fire-damp of its terrors, it only remained for him to enlist it into his service. The simple means by which this was effected are as interesting as their results are important.*

Davy had previously arrived at the fact, that wire-gauze might be substituted as air-feeders to the lamp, in the place of his tubes or safety canals; but not until after the lapse of several weeks, did the happy idea of constructing the lamp entirely of wire-gauze occur to him:—the history of this elaborate enquiry affords a striking proof of the inability of the human mind to apprehend simplicities, without a process of complication which works as the grappling machinery of truth.

His original lamp with tubes or canals, as already described, was perfectly safe in the most explosive atmosphere, but its light was necessarily extinguished by it; whereas in the wire-gauze cage, the fire-damp itself continues to burn, and thus to afford to

* "An Account of an Invention for giving Light in explosive mixtures of Fire-damp in Coal Mines, by consuming the Fire-damp." Read before the Royal Society, Jan. 11, 1816.

the miner a useful light, while he is equally secured from the fatal effects of explosion.

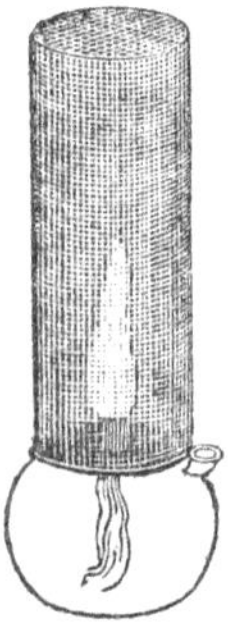

All then required for his guidance and protection in the darkness of the mine, are candles or lamps surrounded by small wire cages, which will at once supply air to the flame, and light to the miner; they may be obtained for a few pence, and be variously modified as circumstances may render necessary.

The reader is here presented with a sketch of the gauze instrument, in its first and simplest form. The original lamp is preserved in the laboratory of the Royal Institution.

Nothing now remained but to ascertain the degree of fineness which the wire-gauze ought to possess, in order to form a secure barrier against the passage of flame. For this purpose, Davy placed his lighted lamps in a glass receiver, through which there was a current of atmospherical air, and by means of a gasometer filled with coal gas, he made the current of air which passed into the lamp more or less explosive, and caused it to change rapidly or slowly at pleasure, so as to produce all possible varieties of inflammable and explosive mixtures; and he found that iron wire-gauze composed of wires from one-fortieth to one-sixtieth of an inch in diameter, and containing twenty-eight wires, or seven hundred and eighty-four apertures to the inch, was safe under all circumstances in atmospheres of this kind; and he consequently employed that material in guarding lamps for the coal mines, where, in

January 1816, they were immediately adopted, and have long been in general use.

Obervations upon them in their working state, and upon the circumstances to which they are exposed, have led to a few improvements or alterations, merely connected with the modes of increasing light or diminishing heat, which were obvious from the original construction.

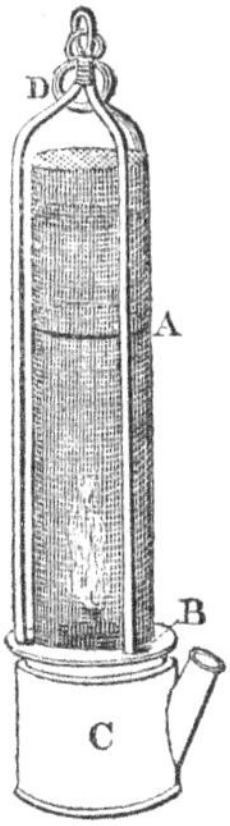

The annexed woodcut represents the lamp which is in present use. A is a cylinder of wire-gauze, with a double top, securely and carefully fastened, by doubling over, to the brass rim B, which screws on to the lamp C. The whole is protected by strong iron supports D, to which a ring is affixed for the convenience of carrying it.

In a paper read before the Royal Society, on the 23rd of January 1817, entitled, "Some new Experiments and Observations on the Combustion of Gaseous Mixtures, with an Account of a method of preserving a continued Light in mixtures of inflammable Gases and Air without Flame," Sir H. Davy announces the application of a principle which he had discovered in the progress of his researches for increasing the utility of the Safety-lamp, and which, a century ago, would have unquestionably exposed its author to the charge of witchcraft.

Having ascertained that the temperature of flame is infinitely higher than that necessary for the ignition of solid bodies, it appeared to him probable that, in certain combinations of gaseous bodies, although the increase of temperature might not be sufficient to render the gaseous matters themselves luminous, they might nevertheless be adequate to ignite solid matters exposed to them. During his experiments on this subject, he was led to the discovery of the curious phenomenon of slow combustion without flame. He observes, that there cannot be a better mode of illustrating the fact, than by an experiment on the vapour of ether or of alcohol. Let a few coils of wire of platinum of the one-sixtieth or one-seventieth of an inch be heated by a hot poker or candle, and let it be brought into the glass; it will presently become glowing, almost white hot, and will continue so, as long as a sufficient quantity of vapour and of air remain in the glass.*

This experiment on the slow combustion of ether is accompanied with the formation of a peculiar acrid and volatile substance possessed of acid properties, which has been particularly examined by

* This principle has been applied for constructing what has been termed the *Aphlogistic Lamp*, which is formed by placing a small coil of platinum wire round the wick of a common spirit lamp. When the lamp, after being lighted for a few moments, is blown out, the platinum wire continues to glow for several hours, as long as there is a supply of spirit of wine, and to give light enough to read by; and sometimes the heat produced is sufficient to rekindle the lamp spontaneously. The same phenomena are produced by the vapour of camphor; and an aromatic fumigating lamp has lately been advertised for sale, which is no other than the contrivance above described; and it is evident that, if the spirit be impregnated with fragrant principles, an aromatic vinegar will be developed during its slow combustion, and diffused in fumes through the apartment.

Mr. Daniell, who, having at first regarded it as a new acid, proposed for it the name of *Lampic* acid, in allusion to the researches which led to its discovery; he has, however, since ascertained that its acidity is owing to the acetic acid, which is combined with some compound of carbon and hydrogen, different both from ether and alcohol.

The phenomena of slow combustion, as exhibited in certain states of the mine, by the Safety-lamp, are highly curious and interesting.

By suspending some coils of fine wire of platinum* above the wick of his lamp, the miner will be supplied with light in mixtures of fire-damp no longer explosive; for should his flame be extinguished by the quantity of fire-damp, the little coil of platinum will begin to glow with a light sufficiently bright to guide him in what would otherwise be impenetrable darkness, and to lead him into a purer atmosphere, when the heat thus increased will very frequently be sufficient to rekindle his lamp!

In this case it will be readily perceived, that the combustion of the fire-damp is continued so slowly, and at so low a temperature, as not to be adequate to that ignition of gaseous matter which constitutes flame, although it excites a temperature sufficient to render platinum wire luminous.

* Sir Humphry Davy attempted to produce the phenomena with various other metals, but he only succeeded with platinum and palladium; these bodies have low conducting powers, and small capacities for heat, in comparison with other metals, which seem to be the causes of their producing, continuing, and rendering sensible, these slow combustions.

Sir Humphry Davy observes, that there never can be any danger with respect to respiration, whenever the wire continues ignited; for even this phenomenon ceases when the foul air forms about two-fifths of the volume of the atmosphere.

The experiment, as originally performed by the illustrious chemist, is so interesting and instructive, that I shall here relate it in his own words.

"I introduced into a wire-gauze Safe lamp a small cage made of fine wire of platinum of one-seventieth of an inch in thickness, and fixed it by means of a thick wire of the same metal about two inches above the wick which was lighted. I placed the whole apparatus in a large receiver, in which, by means of a gas-holder, the air could be contaminated to any extent with coal gas. As soon as there was a slight admixture of coal gas, the platinum became ignited; the ignition continued to increase till the flame of the wick was extinguished, and till the whole cylinder became filled with flame; it then diminished. When the quantity of coal gas was increased, so as to extinguish the flame, at the moment of the extinction the cage of platinum became white hot, and presented a most brilliant light. By increasing the quantity of the coal gas still farther, the ignition of the platinum became less vivid: when its light was barely sensible, small quantities of air were admitted, its heat speedily increased; and by regulating the admission of coal gas and air, it again became white hot, and soon after lighted the flame in the cylinder, which as usual, by the addition of more atmospherical air, rekindled the flame of the wick."

I have thus related, somewhat in detail, the history of a discovery, which, whether considered in relation to its scientific importance, or to its great practical value, must be regarded as one of the most splendid triumphs of human genius. It was the fruit of elaborate experiment and close induction; chance, or accident, which comes in for so large a share of the credit of human invention, has no claims to prefer upon this occasion, step by step, may he be followed throughout the whole progress of his research, and so obviously does the discovery of each new fact spring from those that preceded it, that we never for a moment lose sight of our philosopher, but keep pace with him during the whole of his curious enquiry.

He commenced, as we have seen, with ascertaining the degree of combustibility of the fire-damp, and the limits in which the proportions of atmospheric air and carburetted hydrogen can be combined, so as to afford an explosive mixture. He was then led to examine the effects of the admixture of azote and carbonic acid gas; and the result of those experiments furnished him with the basis of his first plan of security. His next step was to enquire whether explosions of gas would pass through tubes; and on finding that this did not happen, if the tubes were of certain lengths and diameters, he proceeded to examine the limits of such conditions, and by shortening the tubes, diminishing their diameters, and multiplying their number, he at length arrived at the conclusion, that a simple tissue of wire-gauze afforded all the means of perfect security; and he constructed a lamp, which has been truly declared to be as marvellous in its operation, as the storied lamp of Aladdin, realizing its fabled powers of conducting in safety, through "fiends of combustion," to the hidden treasures of the earth. We behold a power which, in its effects, seemed to emulate the violence of the volcano and the earthquake, at once restrained by an almost invisible and impalpable barrier of network—we behold, as it were, the dæmon of fire taken captive by Science, and ministering to the convenience of the miner, while harmlessly fluttering in an iron cage.

And yet, wonderful as the phenomenon may appear, his experiments and reasonings have demonstrated, that the interruption of flame by solid tissues permeable to light and air, depends upon no recondite or mysterious cause, but simply upon their cooling powers, which must always be proportional to the smallness of the mesh, and the mass of the metal.

When it is remembered that the security thus conferred upon the labouring community, is not merely the privilege of the age in which the discovery was effected, but must be extended to future times, and continue to preserve human life as long as coal is dug from our mines, can there be found in the whole compass of art or science, an invention more useful and glorious?

The wire-gauze lamp has now been several years extensively used in the mines, and the most satisfactory and unequivocal testimonies have been published of the complete security which it affords. They have amongst the miners obtained the name of

Davys; and such is the confidence of the work men in their efficacy, that by their aid they enter the most explosive atmospheres, and explore the most remote caverns, without the least dread of their old enemy the *fire-damp*.

Into the mines of foreign countries the Safety-lamp has been introduced with similar success; and the illustrious discoverer has been repeatedly gratified by accounts of the enthusiasm with which his invention has been adopted in various parts of Europe.*

* A pamphlet appeared at Mons, in the year 1818, on the explosions that occur in coal mines, and on the means of preventing them by Davy's Safety-lamp. It was published under the direction of the Chamber of Commerce and Manufactures of Mons, accompanied by notes, and by the results of a series of experiments that had been conducted by M. Gossart, President of the Chamber. The province of Hainault is said to be richer in coal mines than any other part of the Continent of Europe, and to have no less than one hundred thousand persons employed in the working them. The same kind of dangerous accidents occurred in these mines as in those of the North of England, and various expedients had been adopted for their prevention, which, however, availed but little in obviating them. "All the precautions," observe the reporters, "which had been hitherto known or practised, had not been able to preserve the unfortunate miners from the terrible effects of explosion. It is therefore an inappreciable benefit which we confer by making known the equally simple and infallible method of preventing these accidents, which has been discovered by the celebrated Humphry Davy."

M. Gossart gives an ample and accurate detail of the properties of the explosive gas, and confirms the truth of Davy's experiments, by which the high temperature necessary for its inflammation, and the consequent means of preventing it, by reducing that temperature, as effected during its passage through wire-gauze, are clearly demonstrated.

The lamp appears from this report to have been as useful in

Nor is the utility of this invention limited to the operations of mining. In gas manufactories, spirit warehouses,* or druggists' laboratories, and in various other situations, where the existence of an explosive atmosphere † may expose persons to danger, the Safety-lamp may be advantageously used; and as science proceeds in multiplying the resources of art, this instrument will no doubt be found capable of many new applications.

By the permission of the President and Council of the Royal Society, several accounts of these researches, and of the invention and use of the Safety-lamp were printed, and circulated through the coal districts.

It might have been fairly expected that, in a district which had been so continually and so awfully

the mines of Flanders as in those of England. The Pamphlet is a valuable document, inasmuch as it affords an independent proof of the security of the instrument, and displays the high sense of obligation which foreign nations entertain to Sir Humphry Davy for his invention.

* The danger of carrying a naked light into an atmosphere impregnated with the fumes of spirit was awfully exemplified by the loss of the Kent East Indiaman, by fire, in the Bay of Biscay, on the 1st of March 1825.

† In cases where there is any suspicion of accumulations of carburetted hydrogen from the leakage of gas pipes, or from other sources, the safety-lamp should always be employed. A terrible accident occurred some years since at Woolwich, from a room filled with the vapour of coal-tar, for the purpose of drying and seasoning timber intended for ship-building. As the combustion arose from the flame issuing through the flue, which ran along the apartment, at the moment the damper was applied at the top of the building, it is evident that, had a wire-gauze guard been used, the accident could not have occurred. The house was completely demolished, and nine persons were unfortunately killed.

visited by explosions, against which no human foresight had as yet been able to provide a remedy, the disinterested services of the greatest chemist of the age would at least have been received without a dissentient voice, and that his invention of security would have escaped the common fate of all great discoveries, and been accepted with every homage of respect and gratitude; but the inventor of the Safety-lamp was doomed to encounter a bitter hostility from persons whom a spirit of rivalry, or a feeling of hopeless emulation, had cemented into a faction.

From the period of the first announcement of the Safety-lamp, a prejudice against its use was industriously circulated amongst the miners; and some persons even maintained the monstrous proposition, that any protection against the explosions of fire-damp would injure more than it could serve the collier, by inducing him to resume abandoned works, and thus continually to inhale a noxious atmosphere.

The utility of the lamp having been established, in spite of every opposition, the claims of Sir H. Davy to its invention were next publicly challenged.

It will hereafter be scarcely believed that an invention so eminently philosophic, and which could never have been derived but from the sterling treasury of science, should have been claimed in behalf of an engine-wright of Killingworth, of the name of Stephenson—a person not even professing a knowledge of the elements of chemistry. As the controversy to which this claim gave birth has long since subsided, I would willingly have treated it as a passing cloud, had not its shadow remained. The circumstances, however, of the transaction stand recorded in the Magazines of the day, and the biographer of Davy would compromise his rights, by omitting to notice the attempts that have been made to invalidate them.

The claims which were made for the priority of Mr. Stephenson's invention of the Safety-lamp were urged in several communications in the Newcastle Courant. It has been said in reply, that if dates were taken as evidence, not merely of priority, but of originality of invention, it must follow that Mr. Stephenson's lamp was derived from that of Sir H. Davy. With regard to the first of Stephenson's lamps, the only one upon which the shadow even of a claim can be founded, it is unnecessary for the friends of truth to adopt such a line of defence; indeed, after a deliberate examination of all that has been published on the subject, I am very willing to believe that Mr. Stephenson did construct the lamp which dates its origin from the 21st of October 1815, without any previous knowledge of the conclusions at which Davy had arrived; for it was first announced to the Coal-trade by Mr. R. Lambert on the 3rd of November, to the very meeting at which Sir H. Davy's private letter was inadvertently read. —But what were the principles, and what the construction of this lamp?

It would appear that Mr. Stephenson had entertained some vague notion of the practicability of consuming the fire-damp as fast as it entered the lamp, and that if admitted only in small quantities, it would not explode the surrounding atmosphere: for effecting this object, he constructed a lamp with an orifice, over which was placed a slide, by the movement of which the opening could be enlarged, or diminished, and the volume of fire-damp to be admitted into the lamp regulated according to circumstances. Now such a lamp could be nothing else than an exploding lamp; for to make it burn in common air, the orifice must have been so wide that, on going into an explosive atmosphere, the combustion in the interior could not have failed to pass it, and to have exploded the mine. Here then is a *safety*-lamp, which as long as it is safe, will not burn, and the moment it begins to burn, it becomes unsafe!

The testimonies in favour of the security afforded by this lamp were evidently procured from persons who were not only ignorant of the principles of its construction, but of the methods to be pursued for ascertaining its safety. I am surely justified in such a statement, when, instead of an *explosive mixture*, I find them throwing in *pure fire-damp*, which will always extinguish flame, whether burning in a safe or unsafe lamp.

The importance and utility of Davy's lamp having been completely established by the severest ordeals, the general gratitude of the country began more publicly to display itself, and a very strong feeling prevailed, that some tribute of respect should be paid to its inventor; in accordance with which, a deputation of the Coal-owners of the rivers Tyne and Wear, and of the ports of Hartley and Blyth, requested the honour of an interview with Sir H. Davy; upon which occasion they presented him with the following letter, containing an expression of the thanks of the Coal-owners.

TO SIR HUMPHRY DAVY, LL.D. &c.

SIR, Newcastle, March 25, 1816.

As chairman of the general meeting of proprietors of coal-mines upon the rivers Tyne and Wear, held in the Assembly-rooms at Newcastle, on the 18th instant, I was requested to express to you their united thanks and approbation for the great and important discovery of your Safety-lamp for exploring mines charged with inflammable gas, which they consider admirably calculated to obviate those dreadful calamities, and the lamentable sacrifice of human life, which of late years have so frequently occurred in the mines of this country.

They are most powerfully impressed with admiration and gratitude towards the splendid talents and brilliant acquirements that have achieved so momentous and important a discovery, unparalleled in the history of mining, and not surpassed by any discovery of the present age; and they hope that, whilst the tribute of applause and glory is showered down upon those who invent the weapons of destruction, this great and unrivalled discovery for preserving the lives of our fellow-creatures, will be rewarded by some mark of national distinction and honour. I am, Sir,

Your most obedient humble Servant,

GEORGE WALDIE, Chairman.

A plan, however, was under consideration for recording the admiration and gratitude of the Coal-owners, by a more permanent and solid memorial. The nature of this proposition will be best disclosed by inserting the following letter from Mr. Buddle.

TO THE REV. DR. GRAY.

SIR, Wall's-End Colliery, August 27, 1816.

As I know that you feel much interest in all matters relating to Sir H. Davy's Safety-lamp, I trust you will excuse the liberty I take in informing you, that the Committee of the Tyne, approving highly of the suggestion, that some mark of acknowledgment and respect should be presented to Sir Humphry by the Coal-trade of this country, for the happy invention of his lamp, have convened a general meeting of Coal-owners, to be holden at my office in Newcastle, on Saturday next the 31st instant, at twelve o'clock, to take the subject into consideration.

I should have sooner informed you of this proposed meeting, had I not been detained in Cumberland until yesterday; but I shall have the pleasure of transmitting to you a copy of its resolutions.

I am sure that you will be gratified to learn that the lamps continue to go on as well as possible. We now have twelve dozen of them in daily use at this place. I have the pleasure to remain, with the greatest respect, Sir,

Your most obedient humble Servant,

JOHN BUDDLE.

TO THE SAME.

SIR, Newcastle, September 7, 1816.

I NOW have the pleasure of sending you a copy of the Resolutions of the general meeting of Coal-owners on the 31st instant, and shall take the liberty of informing you of the future progress of this affair.

Sir Humphry did me the honour yesterday to

accompany me through the workings of a coal-pit at Wall's-End, when I had an opportunity of witnessing several interesting experiments on his Safety-lamp; and I have the satisfaction to add, that I believe he has now advanced it to the highest degree of perfection. I am, respectfully, Sir,

Your humble Servant,

JOHN BUDDLE.

The satisfactory result of this visit Sir Humphry communicated to Mr. Lambton, now Lord Durham; and I shall take this opportunity to state, that for this as well as for several other letters I shall hereafter have occasion to introduce, I am indebted to that noble Lord, through the kind application of my friend Sir Cuthbert Sharp.

TO J. G. LAMBTON, ESQ. M. P.

MY DEAR SIR, Newcastle, September 9, 1816.

SINCE I last had the pleasure of seeing you, I have examined the workings in the Wall's-End collieries by the lamps, and have tried them in various explosive mixtures.

On Sunday, I went with Mr. Buddle to your *blower*, with the single lamps furnished with small tin reflectors. This simple modification rendered them perfectly safe, even in the furious *blow-pipe*, and at the same time increased their light. Nothing could be more satisfactory than all the trials.

I have left a paper in the hands of the Rev. J. Hodgson, which will be printed in a day or two; and I have desired him to send you ten copies, or as many more as you may like to have.

I trust I have now left nothing undone as to the perfect security of the lamps, under every possible circumstance.

I feel highly gratified that it was at your mines I effected the only object that remained to be accomplished—that of guarding against *blowers* meeting fresh currents of air.

I thank you very sincerely for the interest you have taken in the lamps, connected with my efforts to render them applicable in all cases. I remain, &c.

H. DAVY.

On the 19th of October 1816, a letter appeared in the Durham County Advertiser, dated "Gosforth, August 22nd, 1816," in the name of Mr. W. Brandling, in which, alluding to the Resolutions of the Coal-owners of the 31st of August, he expresses a wish that a strict examination should take place previous to the adoption of a measure which might convey a decided opinion to the public, as to the person to whom the invaluable discovery of the Safety-lamp is actually due. "The conviction," says he, "upon my mind is, that Mr. George Stephenson, of Killingworth Colliery, is the person who first discovered and applied the principle upon which safe lamps may be constructed; for, whether the hydrogen gas is admitted through capillary tubes, or through the apertures of wire-gauze, which may be considered as merely the orifices of capillary tubes, does not, as I conceive, in the least affect the principle.

"In the communications I have seen from Sir H. Davy, no dates are mentioned; and it is by a reference to them only that the question can be fairly decided. For the information of the Meeting, therefore, I shall take the liberty of enclosing some which I received from Mr. Stephenson, to the correctness of which, as far as I am concerned, I can bear testimony; at the same time I beg leave to add, that the principle of admitting hydrogen gas in such small detached portions that it would be consumed by combustion,* was, I understand, stated by him to several gentlemen, as the idea he had embraced two months before his lamp was originally constructed."

Mr. Brandling then proceeds to state, that the Killingworth lamp, with a tube to admit the air, and a slide at the bottom of such tube to regulate the quantity to be admitted, was first tried in the Killingworth pits on Saturday October the 21st, 1815; but not being found to burn well, another was ordered the same day with three capillary tubes to admit the air; and on being tried in the mine on the 4th of November following, was found to burn better and to be perfectly *safe.*† On the 17th of November, it was *tried*‡ at Killingworth office with inflammable air before Richard Lambert, Esq.; and on the 24th of the same month, before C. J. Brandling, Esq. and Mr. Murray.

* Granted:—but what connexion has that with the principle of Davy's lamp, or with any *Safety* lamp?

† It could not have been safe.

‡ "Tried"—but how was it tried?—by forcing in *pure* fire-damp, which will extinguish any lamp, instead of exposing the flame to an explosive mixture, which could alone furnish any test of its security.

"On the 30th of November," he says, "a lamp was tried in the mine, in which the air was admitted by means of a double row of small perforations, and found to be perfectly safe, and to burn extremely well."*

At an adjourned Meeting of the Coal-owners, held on the 11th of October 1816, J. G. Lambton, Esq. M.P. in the chair; Mr. William Brandling moved—"That the meeting do adjourn, until, by a comparison of dates, it shall be ascertained whether the merit of the Safety-lamp belongs to Sir Humphry Davy or to Mr. George Stephenson."

On the question being put thereon, THE SAME PASSED IN THE NEGATIVE.

A great number of the Coal-owners, instead of pursuing the idea which had at first been suggested, of a general contribution on the vend, immediately commenced a subscription of individual proprietors of coal-mines; a measure which, it was thought, would express more distinctly and unequivocally the opinion of the trade as to the merit of the invention. The plan is developed in the following letter.

TO THE REVEREND DR. GRAY.

SIR, Wall's-End Colliery, October 27, 1816.

IT is the anxious wish of almost every individual in the trade to compliment Sir Humphry Davy, in that way which may be most grateful to his feelings.

* Very likely: but the reader will please to recollect, that Sir H. Davy had, before this, published an account of his principle of safety by systems of tubes or canals.

It has been suggested that the object will be best attained by substituting an individual (colliery) subscription, instead of the proposed contribution on the vend; and it will at the same time show more distinctly the real opinion of the trade as to the merit of the invention.

This idea was not suggested till yesterday afternoon, and of course there has been but little time to communicate it to the several Coal-owners; but *all* who have heard of the plan approve of it.

To facilitate the business, the committee have formed the annexed scale of contribution.†

* * * * * *

I trust, Sir, you will excuse the trouble which I have given you on this subject; but I am aware that you must feel interested in it; and I hope, Sir, you will allow me to add, that I am fully sensible of the obligation which the Coal trade is under to yourself, for having drawn Sir H. Davy's attention to that particular line of investigation, which has led to the important discovery of the Safety-lamp. I am, Sir, with the greatest respect, your most obedient humble servant,

JOHN BUDDLE.

Some slight alterations were afterwards made in this scheme, in consequence of a wish having been expressed that the Bishop of Durham and the Duke of Northumberland should take the lead in a subscription. The following letter conveys some farther information upon this subject.

† I have not thought it necessary to enumerate the various sums which the different mines were called upon to contribute.

TO THE REVEREND DR. GRAY.

SIR, Newcastle, January 11, 1817.

I HAVE to acknowledge the receipt of your letter of the 9th instant, communicating the intention of the Reverend the Dean and Chapter of Durham, to subscribe fifty guineas towards the plate to be presented to Sir H. Davy, which, together with two hundred guineas from the Coal-owners of the Wear, makes the subscription amount to nearly £1500, and I shall expect some farther subscriptions.

I am sure it will afford you satisfaction to learn that the lamps still continue to give the most gratifying proofs of the advantages resulting from their invention, and that not a single inch of human skin has been lost by fire, wherever they have been used.

Sir Humphry has just made another important improvement in the lamp, by constructing the cylinder of *twisted* wire-gauze. Lamps thus constructed, possess the singular property of not becoming red-hot, under any circumstances of exposure to explosive mixtures, whether urged by a blast, or in a state of rest. I am with great respect, Sir, your most obedient humble servant,

JOHN BUDDLE.

It may be collected from the following letter, that the Committee, in announcing to Sir H. Davy the intended present of plate, delicately sounded him as to the form in which it would be most agreeable to him.

TO N. CLAYTON, ESQ.

SIR, Grosvenor Street, March 23, 1817.

ON my return to town, after an absence of some days, I found the letter of March the 13th, with which you honoured me, at the Royal Institution. I shall not lose a moment in replying to it, and in expressing my grateful feelings for the very flattering communication it contains.

The gentlemen interested in the coal-mines of the two rivers Tyne and Wear, cannot offer me any testimony of their kindness, which I shall not receive with infinite pleasure.

I hardly know how to explain myself on the particular subject of your letter; but as the Committee express themselves satisfied as to the utility of the Safety-lamp, I can only desire that their present, as it is highly honourable to me, should be likewise useful to my friends, and a small social circle, which it would be as a dinner-service for ten or twelve persons.

I wish that even the plate from which I eat should awaken my remembrance of their liberality, and put me in mind of an event which marks one of the happiest periods of my life.

I cannot find any language sufficiently strong to express my thanks to the gentlemen for the manner in which they have distinguished my exertions in their cause, and in the cause of humanity. I have the honour to remain, &c. H. DAVY.

To revert once again to the faction—for such I must denominate it—which, in opposition to the most unequivocal evidence, continued to support the unjust claims of Mr. Stephenson; it would appear from various letters in my possession, that the feelings of Davy were greatly exasperated by this ungenerous conduct.

I shall introduce one of these letters, playful in the midst of its wrath, addressed to Mr. Lambton, the friend* of his youth, and the manly and kind supporter of his scientific character, in the hour of persecution.

TO J. G. LAMBTON, ESQ. M.P.

MY DEAR SIR, Queen Square, Bath, Oct. 29, 1816.

THE severe indisposition of my wife has altered my plans. Your letter slowly followed me here.

Mr. ——— is one of the persons who, after I had advanced a principle of security for a lamp, came upon the ground to endeavour to jockey me. I was not looking to a prize, I merely came forward to show an animal, the breed of which might be useful, when Mr. ——, Dr. ——, &c. brought their sorry jades, which had never before been seen or heard of, to kick at my blood mare.

I never heard a word of George Stephenson and his lamps till six weeks after my principle of security had been published; and the general impression of the scientific men in London, which is confirmed by what I heard at Newcastle, is, that Stephenson had some loose idea floating in his mind, which he had unsuccessfully attempted to put in practice till after my labours were made known;—then, he made

* It will be remembered that they resided together in the house of Dr. Beddoes. See page 59, vol. i. of these memoirs. In the library at Lambton, there is a good portrait of Sir Humphry.

something like a safe lamp, except that it is not *safe*, for the apertures below are four times, and those above, twenty times too large; but, even if Stephenson's plans had not been posterior to my principles, still there is no analogy between his glass exploding machine, and my metallic tissue, permeable to light and air, and impermeable to flame.

I am very glad that you attended the meeting; your conduct at no very distant period will be contrasted with that of some great coal-proprietors, who find reasons for their indifference, as to a benefit conferred upon them, in insinuations respecting the claims of Dr. Clanny, Mr. Stephenson, and others.

Where men resolve to be ungrateful, it is natural that they should be illiberal; and illiberality often hardens into malignity.

I shall receive any present of plate under your auspices, and those of the Committee over which you preside, with peculiar satisfaction. It will prove to me that my labours have not been disregarded by men of whose good opinion I am proud.

I hope you will not blame me for not taking any notice of the attacks of my enemies in the North. I have no desire to go out of my way to crush gnats that buzz at a distance, and do not bite me, or to quarrel with persons who shoot arrows at the moon, and believe, because they have for an instant intercepted a portion of her light, that they have hit their mark. I am sensible to the circumstances under which you attended the meeting.

I offer you my sincere congratulations, and ardent wishes that you may enjoy all possible happiness.

Believe me, &c. H. Davy.

On the 13th of September 1817, Sir Humphry Davy being expected to pass through Newcastle on his return from Scotland, preparations were made, and notice given of a dinner which it was proposed should take place on the 25th instant, for the purpose of presenting to the illustrious philosopher the service of plate which had been prepared for his acceptance.

Upon this gratifying occasion, a very large party assembled at the Queen's Head, consisting of a numerous and respectable body of Coal-owners, and such other gentlemen as had interested themselves during the progress of the investigation, or taken an active part in promoting the introduction of the lamp into the mines.

After the dinner had concluded, and certain toasts of form had been drunk, Mr. Lambton, who filled the chair on the occasion, rose, and on presenting the service of plate to the illustrious guest, addressed him, in a tone of great animation and feeling, in nearly the following terms:

"SIR HUMPHRY,—It now becomes my duty to fulfill the object of the meeting, in presenting to you this service of plate, from the Coal-owners of the Tyne and Wear, as a testimony of their gratitude for the services you have rendered to them and to humanity.

"Your brilliant genius, which has been so long employed in an unparalleled manner, in extending the boundaries of chemical knowledge, never accomplished a higher object, nor obtained a nobler triumph.

"You had to contend with an element of destruction which seemed uncontrollable by human power; which not only rendered the property of the coal-owner insecure, but kept him in perpetual alarm for the safety of the intrepid miner in his service, and often exhibited to him the most appalling scenes of death, and heart-sickening misery.

"You have increased the value of an important branch of productive industry; and, what is of infinitely greater importance, you have contributed to the lives and persons of multitudes of your fellow-creatures.

"It is now nearly two years that your Safety-lamp has been used by hundreds of miners in the most dangerous recesses of the earth, and under the most trying circumstances. Not a single failure has occurred—its absolute security is demonstrated. I have, indeed, deeply to lament more than one catastrophe, produced by fool-hardiness and ignorance, in neglecting to use the safeguard you have supplied; but these dreadful accidents even, if possible, exalt its importance.

"If your fame had needed any thing to make it immortal, this discovery alone would have carried it down to future ages, and connected it with benefits and blessings.

"Receive, Sir Humphry, this permanent memorial of our profound respect and high admiration—a testimony, we trust, equally honourable to you and to us. We hope you will have as much pleasure in receiving, as we feel in offering it. Long may you live to use it—long may you live to pursue your splendid career of scientific discovery, and to give new claims to the gratitude and praise of the world!"

Sir Humphry having received the plate, replied as follows:

"GENTLEMEN,—I feel it impossible to reply, in an appropriate manner, to the very eloquent and flattering address of your distinguished Chairman. Eloquence, or even accuracy of language, is incompatible with strong feeling; and on an occasion like the present, you will give me credit for no small degree of emotion.

"I have been informed that my labours have been useful to an important branch of human industry connected with our arts, our manufactures, commerce, and national wealth. To learn this from such practical authority is the highest gratification to a person whose ardent desire has always been to apply science to purposes of utility.

"It has been also stated, that the invention which you are this day so highly honouring, has been subservient to the preservation of the lives and persons of a most useful and laborious class of men: this, coming from your own knowledge, founded upon such ample experience, affords me a pleasure still more exalted—for the highest ambition of my life has been to deserve the name of a friend to humanity.

"To crown all, you have, as it were, embodied these sentiments in a permanent and magnificent memorial of your good opinion. I can make only imperfect and inadequate efforts to thank you.

"Under all circumstances of my future life, the recollection of this day will warm my heart; and this noble expression of your kindness will awaken my gratitude to the latest moment of my existence."

Sir Humphry having sat down, and the cheering of the company subsided, the Chairman proposed the health of the illustrious Chemist, in three times three.

"Gentlemen," said Sir Humphry, "I am overpowered by these reiterated proofs of your approbation. You have overrated my merits. My success in your cause must be attributed to my having followed the path of experiment and induction discovered by philosophers who have preceded me: willingly would I divide your plaudits with other men of science, and claim much for the general glory of scientific discovery in a long course of ages.

"Gentlemen, I might dwell at some length upon the great increase of wealth and power to the country, within the last half century, by scientific invention, which never could have existed without coal-mines:—I shall refer only to the improvement in the potteries, to the steam-engine, and to the discovery of the gas lights.

"What an immense impulse has the steam-engine given to the arts and manufactures! How much has it diminished labour, and increased the real strength of the country, far beyond a mere increase of population! By giving facilities to a number of other inventions, it has produced even a moral effect in rendering capital necessary for the perfection of labour, credit essential to capital, and ingenuity and mental energy a secure and dignified species of property.

"Science, Gentlemen, is of infinitely more importance to a state than may at first sight appear possible; for no source of wealth and power can be entirely independent of it; and no class of men are so well able to appreciate its advantages as that to which I am now addressing myself. You have not only derived from it the means of raising your subterraneous wealth, but those also of rendering it available to the public.

"Science alone has made pit-coal such an instrument in the hands of the chemist and mechanic; it has made the elements of fire and water perform operations which formerly demanded human labour, and it has converted the productions of the earth into a thousand new forms of use and beauty.

"Gentlemen, allow me to observe, in conclusion, that it was in pursuing those methods of analogy and experiment, by which mystery had become science, that I was fortunately led to the invention of the Safety-lamp. The whole progress of my researches has been registered in the Transactions of the Royal Society, in papers which that illustrious body has honoured by their biennial medal;* in which I can conscientiously assert, that I have gratefully acknowledged even the slightest hints or offers of assistance which I have received during their composition.

"I state this, Gentlemen, not from vain-glory, but on account of certain calumnious insinuations which have arisen—not in the scientific world, for to that the whole progress of my researches is well known, but in a colliery. I must ever treat these insinuations with contempt; and after the honest indignation which has been expressed against them

* The Rumford Medal, to be hereafter noticed.

by the Coal-owners in general, I cannot feel any anxiety on the subject, nor should I have referred to it at all, did I not believe that the very persons amongst whom these insinuations originated, were extensively benefited by, and were constantly using the invention they would seek to disparage. I could never have expected that such persons would have engaged their respectable connexions in mean attempts to impeach the originality of a discovery, given to them in the most disinterested manner, and for which no return was required but an honest acknowledgment of the benefit, founded upon truth and justice.

"I do not envy them their feelings, particularly at the present moment: I do not wish to enquire into their motives: I do hope, however, that their conduct has been prompted by ignorance rather than by malevolence, by misapprehension rather than by ingratitude.

"It was a new circumstance to me, that attempts to preserve human life, and to prevent human misery, should create hostile feelings in persons who professed to have similar objects in view.

"Gentlemen, I have had some opposition, much labour, and more anxiety, during the course of these researches; but had the opposition, the labour, and the anxiety been a thousand times as great, the events of this day would have been more than a compensation."

Sir Humphry, after drinking the health and happiness of the company, proposed as a sentiment—"Prosperity to the Coal-trade."

The healths of the Duke of Northumberland, the

Bishop of Durham, and the Reverend Dr. Gray, were drunk in succession.

At ten o'clock, Sir Humphry, accompanied by the chairman, retired amidst the enthusiastic plaudits of a meeting, the object of which being one of convivial benevolence, the effect was that of unclouded hilarity."

The party which had supported the claims of Mr. Stephenson had also their meeting; and it was held on the 1st of November. At this meeting it was resolved, "That it was the opinion of the persons present, that Mr. G. Stephenson having discovered the fact, that explosions of hydrogen gas will not pass through tubes and apertures of small dimensions, and having been the first to apply the principle to the construction of a Safety-lamp, is entitled to some reward."

A committee was accordingly formed to carry this resolution into effect, at the head of which was placed the name of the Earl of Strathmore.

The respectable body of Coal-owners, under whose auspices the invention of Sir Humphry Davy had been introduced and rewarded, felt that they owed it to their own characters to repel assertions which amounted to a charge against themselves of ingratitude and injustice: a general meeting was accordingly summoned, at the Assembly-rooms in Newcastle, on the 26th of November 1817, J. G. Lambton, Esq. M. P. in the chair—when it was resolved,

"That the Resolutions passed at the Meeting of the friends of Mr. G. Stephenson on the 5th instant, impugn the justice and propriety of the proceedings of a meeting of the Coal-trade on the 31st of August 1816:

"That the present meeting, therefore, feel themselves called upon, as an act of justice to the character of their great and disinterested benefactor, Sir Humphry Davy, and as a proof that the Coal-trade of the North in no way sanctions the resolutions of Mr. Stephenson's friends, to state their decided conviction, that the merit of having discovered the fact, that explosions of fire-damp will not pass through tubes and apertures of small dimensions, and of having applied that principle to the construction of a Safety-lamp, *belongs to Sir Humphry Davy alone.*

"That this meeting is also decidedly of opinion, from the evidence produced in various publications by Mr. George Stephenson and his friends, subsequently to the meeting of the Coal-trade which was held on the 18th of March 1816, as well as from the documents which have been read at this meeting, that Mr. Stephenson *did not* discover the fact, that explosions of fire-damp will not pass through tubes and apertures of small dimensions; and that he *did not* apply that principle to the construction of a Safety-lamp; and that the latest lamps made by Mr. Stephenson are evident imitations of those of Sir Humphry Davy, and that, even with that advantage, they are so imperfectly constructed as to be actually unsafe.

"That the above resolutions be published thrice in the Newcastle papers, and in the Courier, Morning Chronicle, and Edinburgh Courant; and that printed copies thereof be sent to the Lords Lieutenants of the two counties, to the Lord Bishop of Durham, and to the principal owners and lessors of collieries upon the Tyne and Wear."

The following letter from Sir Humphry Davy announces the farther measures which he also had thought proper to pursue, in order to counteract the impression which the meeting of Mr. Stephenson's friends might have produced on the less informed part of the public.

TO J. G. LAMBTON, ESQ. M. P.

MY DEAR SIR, November 21, 1817.

I SHALL send off by this post a copy of the resolutions, which will appear to-morrow in the Chronicle and Courier.

The men of science who have signed these resolutions are the first chemists and natural philosophers of the country, with the President of the Royal Society, the most illustrious body in Europe, at their head.

It is disagreeable to be thus obliged to use artillery for the destruction of bats and owls; but it was necessary that something should be done.

The Messrs. —— have for a long time been endeavouring to destroy my peace of mind; my offence being that of conferring a benefit.

The only persons I knew in Newcastle, before I gave the Safety-lamp to the Coal-owners, were Dr. Headlam and Mr. Bigge, so that friends I had none; and the few persons with whom I had a slight acquaintance, and who were civil to me before I discovered the Safety-lamp, became my enemies. It requires a deep metaphysician to explain this—Can it be that I did not make them the medium of communication to the colliers?—But I quit a subject to which I have no desire to return, and shall only recollect that day when your eloquence touched my feelings more than it flattered my self-love.

Believe me, &c. &c.

H. DAVY.

The following are the Resolutions of a Meeting adverted to in the preceding letter, and which was held "for considering the Facts relating to the Discovery of the Lamp of Safety."

Soho Square, Nov. 20, 1817.

"An advertisement having been inserted in the Newcastle Courant, of Saturday, November 7, 1817, purporting to contain the Resolutions of 'A Meeting held for the purpose of remunerating Mr. George Stephenson, for the valuable service he has rendered mankind by the invention of his Safety-lamp, which is calculated for the preservation of human life in situations of the greatest danger,'

"We have considered the evidence produced in various publications by Mr. Stephenson and his friends, in support of his claims; and having examined his lamps, and enquired into their effects in explosive mixtures, are clearly of opinion—

"First,—That Mr. George Stephenson *is not* the author of the discovery of the fact, that an explosion of inflammable gas will not pass through tubes and apertures of small dimensions.

"Secondly,—That Mr. George Stephenson *was not* the first to apply that principle to the construction of a Safety-lamp, none of the lamps which he made in the year 1815 having been safe, and there being no evidence even of their having been made upon that principle.

"Thirdly,—That Sir Humphry Davy not only discovered, independently of all others, and without any knowledge of the unpublished experiments of the late Mr. Tennant on Flame, the principle of the non-communication of explosions through small apertures, but that he has also the sole merit of having first applied it to the very important purpose of a Safety-lamp, which has evidently been imitated in the latest lamps of Mr. George Stephenson.

(Signed) "Joseph Banks, P.R.S.
"William Thomas Brande,
"Charles Hatchett,
"William Hyde Wollaston."

Thus terminated a controversy, the discussion of which, I am well aware, many of my readers will consider as having been protracted to a tedious, and perhaps to an unnecessary extent; but the biographer had no alternative. In passing it by without a notice, he would have violated his faith to the public, have given a tacit acknowledgment of the claims of Stephenson, and, in his judgment, have committed an act of gross injustice to the illustrious subject of his history; while by giving only an abridged statement, he would have furnished a pretext for doubt, and an opportunity for malevolence.

It is due also to Sir Humphry Davy to observe, that had he practised more reserve in the communication of his results, the spirit of rivalry would have expired without a struggle,—for it derived its only support and power from the generosity of its victim. Had he secured for himself the advantages of his invention by patent, he might have realized wealth to almost any extent; but to barter the products of his intellectual exertions for pecuniary profit, was a course wholly at variance with every feeling of Davy's mind; and we therefore find him, in the advancement, as at the commencement of his fleeting career, spurning the golden apples from his feet, and hastening to the goal for that prize which could alone reward all his labours—the meed of immortal fame.

From a letter dated Newcastle, August 1830, which I had the pleasure to receive from Mr. Buddle, I extract the following interesting passage:—

"In the autumn of 1815, Sir Humphry Davy accompanied me into some of our fiery mines, to *prove* the efficacy of his lamp. Nothing could be more gratifying than the result of those experiments, as they inspired every body with perfect confidence in the security which his invention had afforded.

"Sir Humphry was delighted, and I was overpowered with feelings of gratitude to that great genius which had produced it.

"I felt, however, that he did not contemplate any pecuniary reward; and, in a private conversation, I remonstrated with him on the subject. I said, 'You might as well have secured this invention by a patent, and received your five or ten thousand a-year from it.' The reply of this great and noble-minded man was,—'No, my good friend, I never thought of such a thing; my sole object was to serve the cause of humanity; and, if I have succeeded, I am amply rewarded in the gratifying reflection of having done so.' I expostulated, saying, that his ideas were much too philosophic and refined for the occasion. He replied, 'I have enough for all my views and purposes; more wealth might be troublesome, and distract my attention from those pursuits in which I delight;—more wealth,' he added, 'could not increase either my fame or my happiness. It might, undoubtedly, enable me to put four horses to my carriage; but what would it avail me to have it said that Sir Humphry drives his carriage-and-four?'"

The present Bishop of Bristol, to whom the world is so greatly indebted for having first called the attention of Sir Humphry Davy to the subject of explosions from fire-damp, and who has kindly interested himself in my arduous and anxious undertaking, was desirous to obtain for me the latest accounts with respect to the Safety-lamp, as to the constancy of its use, and the extent of its security; and his Lordship informs me, that having applied to Mr. Buddle and Mr. Fenwick for information upon these points, their answers have been most satisfactory; at the same time, his Lordship transmitted me much valuable information, which was accompanied by the following letter from Mr. Buddle.

TO THE RIGHT REVEREND THE LORD BISHOP OF BRISTOL.

MY LORD, Wall's-End, August 11, 1830.

I HAVE the honour to acknowledge the receipt of your Lordship's letter of yesterday's date. I am glad your Lordship has interested yourself in Doctor Paris's work, and I hope that he will be enabled, through the assistance of Sir Humphry's friends, to do ample justice to the genius and worth of that excellent man.

I should be very happy if any letters of mine could assist Dr. Paris in doing justice to his merits in the invention of the Safety-lamp; and I shall with pleasure submit to your Lordship's better judgment and discretion the selection of such of them as may seem to be conducive to that object.

I do not find that any improvement whatever has been made, either in the principle or construction of the *original lamp*, as presented to us by Sir Humphry. His transcendent genius seems to have anticipated every thing belonging to the subject, and has left nothing more to be done.

I have the honour to be,
My Lord, with great respect,
Your Lordship's most obedient, humble servant,
JOHN BUDDLE.

In consequence of some late reports of accidents in the mines, I requested my friend Sir Cuthbert Sharp to make certain enquiries in the mining districts; and for this purpose, I sent him a string of queries, to which I begged him to obtain answers. These questions were submitted to Mr. Buddle, and they produced the following letter.

TO SIR CUTHBERT SHARP.

Newcastle, August 28, 1830.

MY DEAR SIR CUTHBERT,

I RETURN Dr. Paris's letter, and shall briefly answer his enquiries.

If the Davy lamp was exclusively used, and due care taken in its management, it is certain that few accidents would occur in our coal mines; but the exclusive use of the "*Davy*" is not compatible with the working of many of our mines, in consequence of their not being workable without the aid of gunpowder.

In such mines, where every collier must necessarily fire, on the average, two *shots* a-day, we are exposed to the risk of explosion from the ignition of the gunpowder, even if no naked lights were used in carrying on the ordinary operations of the mine.

This was the case in Jarrow Colliery, at the time the late accident happened. As the use of gunpowder was indispensable, naked lights were generally used, and the accident was occasioned by a '*bag*' of inflammable air forcing out a large block of coal, in the face of a drift, from a fissure in which it had been pent up, perhaps from the Creation, and firing at the first naked light with which it came in contact, after having been diluted down to the combustible point by a due admixture of atmospheric air.

As to the number of old collieries and old workings which have been renovated, and as to the quantity of coal which has been, and will be saved to the public by the invention of the "*Davy*," it is scarcely possible to give an account, or to form an estimate.

In this part of the country, 'Walker's Colliery,' after having been completely worked out, according to the former system, with candles and steel-mills, and after having been abandoned in 1811, was re-opened in 1818 by the aid of the "*Davy*," and has been worked on an extensive scale ever since, and may continue to be worked for an almost indefinite period.

Great part of the formerly relinquished workings of Wall's-end, Willington, Percy-main, Hebburn, Jarrow, Elswick, Benwell, &c. &c., as well as several collieries on the Wear, have been recovered, and are continued in work by the invention of the "*Davy*."

If I had only what you know perfectly well I have not—TIME, I could write a volume on this subject.

I shall shortly, through the medium of a friend, get an important paper on the subject of the "*Davy*," put into Dr. Paris's hands.

Believe me, my dear Sir Cuthbert,

To remain yours very faithfully,

JOHN BUDDLE.

The Bishop of Bristol has placed at my disposal a communication from Mr. Fenwick, a gentleman of much practical ability, which affords additional evidence of the utility of the lamp; from which the following is an extract.

"Sir H. Davy's safety-lamp has afforded much security in the general working of mines, particularly by enabling the coal-owner to work, in several situations, the pillars of coal formerly left therein, which, under the system of working by candles, or open flame, was deemed hazardous and impracticable; and, in consequence, one-sixth part more of coal may be estimated as obtainable from those mines which are subject to hydrogen gas. Also in the working of the pillars of coal, (commonly called the second working,) great advantages and securities, as well as saving of expenses, have resulted from the use of this lamp, not only to the lessees of collieries, inasmuch as more coal is obtained from a given space than before, (particularly in collieries subject to fire-damp,) but also to the lessor of such mines, by their being more productive, and of course more durable than heretofore.

"Another advantage results from the use of this Safety-lamp, and in the working of the pillars in particular. It is found now, through experience, that the changeable state of the atmosphere, which our barometers daily indicate, has a most powerful effect on the noxious air in mines; as, from a sudden change in the atmosphere, indicated by the rapid fall of the mercury in the barometrical tube, a rapid discharge of noxious gas into the workings and excavations of the mine is the consequence, caused by the want of the atmospheric equilibrium:* in which case the mine becomes suddenly surcharged with hydrogen, and if worked by the light of *open flame*, an explosion may take place before the possibility of such a circumstance can even be suspected; but if worked by the Safety-lamp, it is

* This is a very interesting fact, and gives much support to the theory advanced at page 62 of this volume.

only shown by the gas in the lamp becoming a pillar of harmless flame. This circumstance frequently takes place when any atmospheric change causes the mercury in the barometer to sink to twenty-eight inches and a half or thereabouts."

In the year 1825, Sir Humphry Davy had the honour to receive from the Emperor Alexander of Russia, a superb silver gilt vase, standing in a circular tray enriched with medallions. On the cover was a figure, of about sixteen or eighteen inches in height, representing the God of Fire, weeping over his extinguished torch.

The circumstances under which this vase was presented have been communicated to me by Mr. Smirnove, Secretary to the Embassy.

TO J. A. PARIS, M. D.

DEAR SIR, Wigmore Street, May 29, 1830.

IT was in the month of April, or May, 1815, that the late Sir Humphry Davy expressed to Prince, then Count, Lieven, his wish to offer to the Emperor of Russia a model of his Safety-lamp, which he had recently improved, accompanied by an explanatory pamphlet on the subject.

Prince Lieven of course complied with this request; and the Emperor having been pleased to accept it, ordered the Ambassador, in November of the same year, to thank Sir Humphry for it in his Majesty's name, and to assure him how much his Majesty appreciated the merit of an invention, the double effect of which was to favour the progress of more than one branch of industry, and to ensure the safety of persons employed in the coal-mines, against those fatal accidents which had hitherto so frequently occurred. The Ambassador, at the same time, delivered to Sir Humphry a silver-gilt Vase,* in the name of the Emperor, in testimony of the high satisfaction with which that sovereign had been pleased to accept the object in question. I beg you to believe me, with regard and esteem,

Your faithful servant,

JOHN SMIRNOVE.

It is well known to the friends of Davy, that in his conversation as well as in his correspondence, he always dwelt with peculiar satisfaction and delight upon the invention of his Safety-lamp.

Mr. Poole, in a letter lately addressed to me, observes—" How often have I heard him express the satisfaction which this discovery had given him. 'I value it,' said he, 'more than any thing I ever did. It was the result of a great deal of investigation and labour; but, if my directions be only attended to, it will save the lives of thousands of poor labourers. I was never more affected,' he added, 'than by a written address which I received from the working colliers, when I was in the North, thanking me, in behalf of themselves and their families, for the preservation of their lives.' I remember how delighted he was when he showed me the service of plate presented to him by those very men and their employers, as a testimony of their gratitude."

The following letter evinces a similar feeling.

* Of the value of about one hundred and eighty guineas.

TO THOMAS POOLE, ESQ.

MY DEAR POOLE, Queen's Square, Bath, Oct. 29, 1816.

It is very long since any letters have passed between us. The affections and recollections of friendly intercourse are of a very adhesive nature; and I think you will not be displeased at being put in mind that there is an old friend not very far from you, who will be very glad to see you.

Bath does not suit me much, nor should I remain here, but my wife has been indisposed, and the waters seem to benefit her, and promise to render her permanent service, and if that happens, I shall be pleased even with this uninteresting city.

I have seen many countries and nations since we met.

* * * * * *

I have just come from the North of England, where it has pleased Providence to make me an instrument for preserving the lives of some of my fellow-creatures. You, I know, are of that complexion of mind that the civic crown will please you more than even the victor's laurel wreath.

I have a bed, though a small one, at your service; if you can come here for two or three days, I assure you we shall be most happy to see you. We shall remain in Bath about three weeks. I shall be absent for a few days in the beginning of next week, and after that I shall be stationary till the middle of November. Give me a few lines, and say when we may expect you.

I am, my dear Poole, very sincerely yours,

H. Davy.

TO THE SAME.

MY DEAR POOLE, Grosvenor Street, Dec. 3, 1817.

THE late melancholy event * has thrown a gloom over London, and indeed over England. The public feeling is highly creditable to the moral tone of the people.

The loss of a Princess, known only by good qualities, living in a pure and happy state of domestic peace, is in itself affecting; but when it is recollected that two generations of sovereigns of the first people in the world have been lost at the same moment, the event becomes almost an awful one.

I go on always labouring in my vocation. I am now at work on a subject almost as interesting as the last which I undertook. It is too much to hope for the same success; at least I will deserve it.

When you come to town in the spring, which I trust you will do, I shall show you my service of plate. I do not think you will like it the less for the cause of the gift.

I am not sure whether I shall not take a run down to Nether Stowey and the west for a few days, if you encourage me with any hopes of the estate† and of woodcocks. You will fix my plans.

I shall be disengaged between the 15th and Christmas, and shall like to revisit Lymouth, and above all to shake you by the hand.

* The death of the Princess Charlotte.

† He here alludes to an estate in the neighbourhood of Nether Stowey, which he wished to purchase, and about which he had requested Mr. Poole to make enquiries.

Lady D. is in better health than I have ever known her to possess. She begs her kind remembrances.

I am, my dear Poole, most affectionately yours,

H. DAVY.

In a strictly scientific point of view, the most interesting results which have arisen out of the investigation for constructing a Safety-lamp, are perhaps those which have made us better acquainted with the true nature of flame, and the circumstances by which it is modified; and which have led to some practical views connected with the useful arts.

It is, I think, impossible to enter into the details of those curious investigations* which, under the title of "Some Researches on Flame," were communicated to the Royal Society, and read before that body on the 16th of January 1817, without being forcibly struck with the address by which Davy, in the first instance, brought abstract science to promote and extend practical knowledge; and then, as it were by a species of multiplied reflection, applied the new facts thus elicited for the farther extension of speculative truth; which in its wider range became again instrumental in disclosing a fresh store of useful facts. It may be said to have been the power of dexterously combining such methods which constituted the felicity of his genius; for, in general, each of them requires for its successful application a mind of quite a distinct order and construction. Mr. Babbage has very justly observed, that those intellectual qualifications which give birth to new

* A short notice of them first appeared in the third number of the "Journal of Science and the Arts," edited at the Royal Institution.

principles or to new methods, are of quite a different order from those which are necessary for their practical application. Davy furnished the exception that was necessary to make good the rule.

He detects, in the first instance, the general principle of inflammable gas, in a state of combustion, being arrested in its progress by capillary tubes; he next applies it to the construction of a Safety-lamp, and then, by observing the phenomena which this lamp exhibits, is led to novel views respecting the nature and properties of flame.—I shall endeavour to offer a popular view of the curious and interesting truths disclosed by this latter research.

He had observed that, when the coal gas burnt in the iron cage, its colour was pale, and its light feeble; whereas the fact is rendered familiar to us all by the flame of the gas lights, that in the open air carburetted hydrogen burns with great brilliancy. Upon reflecting on the circumstances of the two species of combustion, he was led to believe that the cause of the superiority of the light in the latter case might be owing to a *decomposition* of a part of the gas towards the interior of the flame, where the air was in the smallest quantity; and that the consequent deposition of charcoal might first by its *ignition,* and afterwards by its *combustion,* contribute to this increase of light. A conjecture which he immediately verified by experiment.*

* This theory of Davy is well illustrated by the change produced in the flame of gas-light, when acted upon by the wind, as may be seen during an illumination. The loss of light under these circumstances evidently arises from the more rapid combustion of the gas, by its more complete admixture with air; in consequence of which the decomposition above described does not take place.

The intensity therefore of the light of flames depends principally upon the production and ignition of *solid* matter in combustion, so that heat and light are in this process independent phenomena.

These facts, Davy observes, appear to admit of many applications; in explaining, for instance, the appearance of different flames—in suggesting the means of increasing or diminishing their light, and in deducing from their characters a knowledge of the composition of their constituent parts.

The point of the inner blue flame of a candle or lamp urged by the blow-pipe, where the heat is the greatest and the light the least, is the point where the whole of the charcoal is burnt in its gaseous combinations, without previous ignition. The flames of phosphorus and of zinc in oxygen, and that of potassium in chlorine, afford examples of intensity of light depending upon the production of *fixed* solid matter in combustion; while on the contrary, the feebleness of the light of those flames, in which gaseous and volatile matter is alone produced, is well illustrated by those of hydrogen and sulphur in oxygen, or by that of phosphorus in chlorine.

From such facts, he is inclined to think that the luminous appearance of shooting stars and meteors cannot be owing to any inflammation of gas, but must depend upon the ignition of solid matter. Dr. Halley calculated the height of a meteor at ninety miles, and the great American meteor, which threw down showers of stones, was estimated at only seventeen miles high. The velocity of the motion of such bodies must in all cases be immensely great, and the heat thus produced by the compression of the most rarefied air, Davy thinks, must be sufficient to ignite the mass; and that all the phenomena may be explained by assuming that *falling stars* are small solid bodies moving round the earth in very eccentric orbits, which become ignited only when they pass with immense velocity through the upper regions of the atmosphere, and which, when they contain either combustible or elastic matter, throw out stones with explosion.

By the application of such a principle did he also infer the composition of a body from the character of its flame: thus, says he, Ether, during its combustion, would appear to indicate the presence of *olefiant gas*. Alcohol burns with a flame similar to that of a mixture of carbonic oxide and hydrogen; so that the first is probably a binary compound of olefiant gas and water, and the second of carbonic oxide and hydrogen.

When the proto-chloride of copper is introduced into the flame of a candle or lamp, it affords a peculiar dense and brilliant red light, tinged with green and blue towards the edges, which seems to depend upon the separation of the chlorine from the copper by the hydrogen, and the ignition and combustion of the solid copper and charcoal.

The acknowledged fact of the brightest flames yielding the least heat is easily reconciled, when we learn that the light depends upon fixed matter which carries off the heat. It is equally obvious, that by art we may, for practical purposes, easily modify these phenomena.

In the next place, having observed that wire-gauze cooled down flame beyond its combustible point, he was led to enquire into the nature of pure flame; and he readily demonstrated it to be *gaseous matter heated so highly as to be luminous;* and that the temperature necessary for such an effect was much greater than had been imagined, varying, however, in different cases. The flame of a common lamp he proved, by a very simple experiment, to exceed even the white heat of solid bodies, and which is easily shown by the simple fact of heating a piece of platinum wire over the chimney of an Argand lamp fed with spirit of wine; when it will be seen that air, which is not of sufficient temperature to appear luminous, is still sufficiently hot to impart a white heat to a solid body immersed in it.

The fact of different gaseous bodies requiring different degrees of heat to raise them into flame, was an inference immediately deducible from the phenomena of his *safety gauze.* A tissue of one hundred apertures to the square inch, made of wire of one-sixtieth, will, at common temperatures, intercept the flame of a spirit-lamp, but not that of hydrogen; and, when strongly heated, it will no longer arrest the flame of the spirit-lamp. A tissue which, when red-hot, will not interrupt the flame of hydrogen, will still intercept that of olefiant gas; and a heated tissue, which would communicate explosion from a mixture of olefiant gas and air, will stop an explosion of fire-damp. Fortunately for the success of the Safety-lamp, carburetted hydrogen requires so high a temperature to carry on its combustion, that even metal, when white-hot, is far below it; and hence red-hot gauze, in sufficient quantity, and of the proper degree of fineness, will abstract heat enough from the flame to extinguish it.

The discovery of the high temperature which is necessary for the maintenance of flame, suggested to the philosopher the reason of its extinction under various circumstances. He considers, that the common operation of blowing out a candle principally depends upon the cooling power of the current of air projected into the flame;* and he observes, that the hottest flames are those which are least easily blown out. He farther illustrated this subject by surrounding a very small flame with a ring of *metal*, which had the effect of cooling it so far as to extinguish it; but a ring of *glass*, of similar dimensions and diameter, being a less perfect conductor of heat, produced no such effect.

It had been long known that flame ceased to burn in highly rarefied air; but the degree of rarefaction necessary for this effect had been very differently stated. The cause of the phenomenon was generally supposed to depend upon a deficiency of oxygen.

In the commencement of his enquiry into this subject, Davy observed that the flame of hydrogen gas, the degree of rarefaction and the quantity of air being the same, burnt longer when it issued from a larger than a smaller jet,—a fact the very reverse of that which must have happened had the

* *Quere.* Is this theory correct? May not the effect be mechanical, the appulse of the air separating the flame from the wick.—Upon the principle suggested by Davy, how are we to explain the fact of rekindling the flame by a blast?

flame expired for want of oxygen; he moreover observed, that when the larger jet was used, the point of the glass tube became white-hot, and continued red-hot till the flame was extinguished: he therefore concluded, that the heat communicated to the gas by this tube was the cause of its protracted combustion, and that *flame expired in rarefied air, not for want of nourishment from oxygen, but for want of heat, and that if its temperature could be preserved by some supplementary aid, the flame might be kept burning.* The experiment by which he confirmed this theory was as beautiful as it was satisfactory.

He burnt a piece of camphor in a *glass* tube, under the receiver of an air-pump, so as to make the upper part of the tube red-hot; its inflammation was found to continue when the rarefaction was nine times; but by repeating the experiment in a *metallic* tube, which could not be so considerably heated by it, it ceased after the rarefaction exceeded six times.

It follows then that by artificially imparting heat,* bodies may be made to burn in a rarefied air, when under other circumstances they would be extinguished.

The following may be considered as an *experi-*

* "It is upon this principle that, in the Argand lamp, the Liverpool lamp, and in the best fire places, the increase of effect does not merely depend upon the rapid current of air, but likewise upon the heat preserved by the arrangement of the materials of the chimney, and communicated to the matters entering into inflammation." The art of making a good fire depends also upon the same principle of economising the heat.

mentum crucis, in proof of the fact that combustibility is neither increased nor diminished by rarefaction.

He introduced the flame of hydrogen, in which was inserted a platinum wire, into a receiver of rarefied air, and he found that, as long as the metal remained at a dull red-heat, the flame continued to burn: now it so happens that the temperature, at which platinum approaches a red-heat, is precisely that at which hydrogen inflames under the ordinary pressure of the atmosphere; whence it follows, that its combustibility is not altered by rarefaction.

The same law was found to apply to the flames of other bodies; those requiring the least heat for their combustion always sustaining the greater rarefaction without being extinguished.*

Hitherto he had only considered the effects of rarefaction, when produced by the diminution of pressure; he had next to investigate the phenomena of rarefaction when occasioned by expansion from heat.

The experiments of M. de Grotthus had apparently shown that rarefaction by heat destroys the combustibility of gaseous mixtures; those of Davy, however, proved that it enables them to explode at a lower temperature.

* From a calculation of the ratio in which the density of the atmosphere decreases with its altitude, and from that of the relative combustibility of different bodies, it follows that the taper would be extinguished at a height of between nine and ten miles—hydrogen, between twelve and thirteen—and sulphur, between fifteen and sixteen.

In the progress of this research, while passing mixtures of hydrogen and oxygen through heated tubes, the heat being still below redness, he observed that steam was formed without any combustion.

Here was a slow combination without combustion, as long since observed with respect to hydrogen and chlorine, and oxygen and metals; and he believes that such a phenomenon will happen at certain temperatures with most substances that unite by heat. On trying charcoal, he found that at a temperature which appeared to be a little above the boiling point of quicksilver, it converted oxygen pretty rapidly into carbonic acid, without any luminous appearance, and that, at a dull red-heat, the elements of olefiant gas combined in a similar manner with oxygen, slowly and without explosion.

It occurred to Davy, in the progress of these experiments, that, during this species of slow combination, although the increase of temperature might not be sufficient to render the gaseous matters luminous, or to produce flame, it might still be adequate to ignite solid matters exposed to them. It was while engaged in devising experiments to ascertain this fact, that he was accidentally led to the discovery of the continued ignition of platinum wire, during the slow combination of coal gas with atmospheric air; the circumstances of which have been already related, as well as the curious invention to which the application of the fact gave origin.*

For this and his preceding papers on the subjects

* See page 100 of this volume.

of flame and combustion, the President and Council of the Royal Society adjudged to Sir Humphry Davy the gold and silver medals, on the donation of Count Rumford;* and never, I will venture say, did a society in awarding a prize more faithfully comply with the intentions of its founder.

On the completion of these laborious enquiries, it was thought expedient to give a wider circulation to their results than the publication of them in the Philosophical Transactions was calculated to afford; and Sir Humphry Davy was therefore induced to reprint his principal memoirs, so as to form an octavo

* At the Anniversary of the Royal Society, November 1796, Count Rumford transferred one thousand pounds, Three per Cent. Consols, to the use of the Society, on condition that a premium should be biennially awarded to the author of the most important discovery, or useful invention, made known in any part of Europe during the preceding two years, on the subject of HEAT AND LIGHT. In regard to the form in which this premium was to be conferred, he requested that it might always be given in two medals, struck in the same die, the one of gold, and the other of silver.

Should not any discovery or improvement be made during any terms of years, he directed that the value of the medals should be reserved, and being laid out in the purchase of additional stock, go in augmentation of the capital of this premium.

Medals upon this foundation have been successively voted to Professor Leslie, for his Experiments on Heat, published in his work entitled "An Experimental Enquiry into the Nature and Properties of Heat;"—To Mr. William Murdoch, for his publication "On the Employment of Gas and Coal for the purpose of Illumination;"—to M. Malus, for his discoveries of certain new properties of reflected light;—to Dr. Wells, for his Essay on Dew;—to Sir Humphry Davy, as above stated;—to Dr. Brewster, for his Optical Investigations;—and, lastly, to Mr. Fresnel, for his optical researches.

volume,* which might be accessible to the practical parts of the community.

The enlightened friends of science very reasonably expected that a service of such importance to society as the invention of the Safety-lamp, would have commanded the gratitude of the State, and obtained for its author a high parliamentary reward; nor were there wanting zealous and disinterested persons to urge the claims of the Philosopher: but a Government which had bestowed a splendid pension upon the contriver of an engine* for the destruction of human life, refused to listen to any proposition for the reward of one who had invented a machine for its preservation. It is true that, in consideration of various scientific services, they tardily and inadequately acknowledged the claims of Davy, by bestowing upon him the dignity of Baronetcy ‡—a reward, it must be confessed, that neither displayed any regard to his condition, nor implied the just estimate of his merits. The measure of value, however, enables us to judge of the standard by which the State rates the various services to society; and deeply is it to be lamented that the disproportioned exaltation of military achievement, crowned with the highest honours, depresses respect for science, and raises a false and fruitless object of ambition.

The passion for arms is a relict of barbarity de-

* "On the Safety-lamp for Coal Mines, with some Researches on Flame.—London, 1818."

† Sir William Congreve, in addition to other marks of favour, received a pension of twelve hundred a-year, for the invention of his Rocket; or, in the exact terms of the grant, "for inventions calculated to destroy or annoy the enemy."

‡ He was created a Baronet on the 20th of October, 1818.

rived from the feudal ages; the progress of civilization, and the cultivation of the mind, should have led us to prefer intellectual to physical superiority, and to recognise in the successes of science the chief titles to honour. This reversal of the objects of importance can never be redressed until the aristocracy shall be possessed of a competent share of scientific knowledge, and instructed to appreciate its value. To effect such a change, the system of education so blindly and obstinately continued in our great public schools, must be altered; for minds exclusively applied to classical pursuits, and trained to recognise no other objects of liberal study, are indisposed and indeed disqualified for enquiries ministering to the arts of life, and arrogantly despised for their very connexion with utility. It is in the early ignorance of the rudiments of science that the after negligence of science has its source.

The instances in proof of the extent of the ignorance and indifference I have noted, and of their pernicious effects upon the most important interests of society, especially legislation, and the administration of justice, are abundant. In Parliament, how is a question of science treated? In our courts of law, and criminal investigation, it is lamentable to observe the frequent defeat of justice, arising from erroneous conception, or from the utter absence of the requisite knowledge. In the ordinary affairs of life, we see conspicuous, amongst the dupes of quackery and imposture, those whose stations should imply the best instruction, and whose conduct, unfortunately, has the effect of example.

A contempt far-spreading, and proceeding from

the well-springs of truth, is rapidly rising against this exalted ignorance; the industrious classes of society are daily becoming more imbued with knowledge upon scientific subjects, and the nobility, if they would preserve their superiority in social consideration, must descend to the popular improvement.

Before concluding the present chapter, I must carry back my history to the year 1815, for the purpose of recording a circumstance in the life of Davy, which, while it exemplifies his general love of science, evinces the local attachment he retained for the town of his birth.

In the year 1813, the Geological Society of Cornwall was established at Penzance. Its objects are to cultivate the sciences of Mineralogy and Geology, in a district better calculated perhaps for such pursuits than any other spot in Europe,—to register the new facts which are continually presenting themselves in the mines, and to place upon permanent record, the history of phenomena which had hitherto been entrusted to oral tradition; but, above all, its object was to bring science in alliance with art; to prevent the accidents which had so frequently occurred from explosion in the operation of blasting rocks; and, in short, to render all the resources of speculative truth subservient to the ends of practical improvement.

No sooner had the establishment of so useful an institution been communicated to Davy, than he testified his zeal for its welfare by a handsome donation to its funds; which was followed by a present of a very extensive suite of specimens, illustrative of the volcanic district of Naples, and which had been collected by himself. He also afterwards communicated to the Society a memoir on the Geology of Cornwall, which has been published in the first volume of its Transactions.

In this paper, he discusses several of the more difficult questions connected with the origin of veins.

He first observed the granitic veins, which have called forth so much attention from geologists, about the year 1797; probably before they had excited much scientific notice: he is disposed to regard them as peculiar to the low metalliferous granite and mica formations; he had seen several cases of granite veins near Dublin, in the Isle of Arran, and in other parts of Scotland; he had also observed several instances near Morlaix in Brittany, but he had in vain searched for them in the points of junction of the schist and granite, both in the Maritime, Savoy, Swiss, and Tyrolese Alps, and likewise in the Oriental Pyrenees.

The *serpentine* district of Cornwall, he thinks, has not yet met with the attention it deserves. "I have seen no formation," says he, "in which the nature of serpentine is so distinctly displayed. The true constituent parts of this rock appear to be *resplendent hornblende* and *felspar*; it appears to differ from *sienite* only in the nature of the *hornblende*, and in the chemical composition of its parts, and in being intersected by numerous veins of *steatite* and *calcareous spar*."

The nature and origin of the veins of *steatite* in serpentine, he considers as offering a very curious subject for enquiry. "Were they originally crystallized," he asks, "and the result of chemical deposition? or have they been, as for the most part they are now found, mere mechanical deposits?" He is inclined to the latter opinion. The felspar in serpentine, he observes, is very liable to decomposition, probably from the action of carbonic acid and water on its alkaline, calcareous, and magnesian elements; and its parts washed down by water and deposited in the chasms of the rocks, he thinks would necessarily gain that kind of loose aggregation belonging to steatite.

He had some years before made a rude, comparative analysis of the felspar in serpentine, and of the soap-rock, when he found the same constituents in both of them, except that there was not any alkali or calcareous earth in the latter substance. It is very difficult to conceive, he says, that steatite was originally a crystallized substance which has been since decomposed; for, in that case, it ought to be found in its primitive state in veins which are excluded from the action of air and water; whereas it is easy to account for the hardness of some species of steatite on the former hypothesis; for mere mechanical deposits, when very finely divided, and very slowly made, adhere with a very considerable degree of force. A remarkable instance of this kind occurred to him amongst the chemical preparations of the late Mr. Cavendish, which, on the decease of that illustrious philosopher, had been presented to him by Lord George Cavendish: there was a bottle which had originally contained a solution of silica by potash; the cork, during the lapse of years, had become decayed, and the carbonic acid of the atmosphere had gradually precipitated the earth, so that it was found in a state of solid cohesion; the upper part was as soft as the steatite, but the lower portion was extremely hard, was broken with some difficulty, and presented an appearance similar to that of chalcedony.

In speaking generally of the mineralogical interest of Cornwall, he observes, that "it may be regarded, κατ' εξοχην, as the country of veins; and that it is in veins that the most useful as well as the most valuable minerals generally exist, that the pure specimens are found which serve to determine the mineralogical species, and that the appearances seem most interesting in their connexion with geological theory. Thus veins, which now may be considered in the light of the most valuable cabinets of nature, were once her most active laboratories; and they are equally important to the practical miner, and to the mineralogical philosopher.

With regard to the general conformation of Cornwall, he states it to be in the highest degree curious, and he considers that the facts which it offers are illustrative of many important points of geological theory. "It exhibits very extraordinary instances of rocks broken in almost every direction, but principally from east to west, and filled with veins again broken in, diversified by cross lines, and filled with other veins, and exhibiting marks of various successive phenomena of this kind.

"Respecting the agents that produced the chasms

in the primary strata, and the power by which they were filled with stony and metallic matter, it would be easy to speculate, but very difficult to reason by legitimate philosophical induction."

In the concluding passage, however, he very freely admits his preference for the doctrine of fire.

"It is amongst extinct volcanoes, the surfaces of which have been removed by the action of air and water, and in which the interior parts of strata of lavas are exposed, that the most instructive examples of the operation of slow cooling upon heated masses are to be found. It is difficult to conceive that water could have been the solvent of the different granitic and porphyritic formations; for, in that case, some combinations of water with the pure earths ought to be found in them. Quartz ought to exist in a state of *hydrate*, and Wavellite, not Corundum, ought to be the state of alumina in granite.

"To suppose the primary rocks, in general, to have been produced by the slow cooling of a mass formed by the combustion of the metallic bases of the earths, appears to me the most reasonable hypothesis; yet aqueous agency must not be entirely excluded from our geological views. In many cases of crystallization, even in volcanic countries, this cause operates; thus in Ischia, siliceous *tufas* are formed from hot springs; and in the lake Albula, or the lake of Solfaterra, near Tivoli, crystals of calcareous spar and of sulphur separate from water impregnated with carbonic acid and hepatic gas; and large strata of calcareous rocks, formed evidently in late times by water impregnated with carbonic acid, exist in various parts of Europe. The Travertine marble

(Marmor Tiburtinum) is a production of this kind; and it is of this species of stone that the Coloseum at Rome, and the cathedral of St. Peter, are built. It is likewise employed in the ancient temple of Pæstum, and it rivals in durability, if not in beauty, the primary marble of Paris and Carrara."

CHAPTER XII.

Sir Humphry Davy suggests a chemical method for unrolling the ancient Papyri.—He is encouraged by the Government to pro ceed to Naples for that purpose.—He embarks at Dover.—His experiments on the Rhine, the Danube, the Raab, the Save, the Ironzo, the Po, and the Tiber, in order to explain the formation of mists on rivers and lakes.—His arrival and reception at Naples. —He visits the excavations at Herculaneum.— He concludes that it was overwhelmed by sand and ashes, but had never been exposed to burning matter —He commences his attempt of unrolling the Papyri.—His failure.—He complains of the persons at the head of the department in the Museum.—He analyses the waters of the Baths of Lucca.—His return to England.—Death of Sir Joseph Banks.—He is elected President of the Royal Society —Some remarks on that event.—He visits Penzance.—Is honoured by a public dinner.—Electro-magnetic discoveries of Oersted extended by Davy.—He examines Electrical Phenomena in vacuo.—The results of his experiments questioned.—He enquires into the state of the water, and aëriform matter in the cavities of crystals.—The interesting results of his enquiry confirm the views of the Plutonists.

Our history now proceeds to exhibit Sir Humphry Davy in quite a new field of enquiry;—engaged in investigating, amidst the ruins of Herculaneum, the nature and effects of the volcanic eruption which overwhelmed that city in the reign of Titus; and in attempting, by the resources of modern science, to unfold and to render legible the mouldering archives which have been recovered from its excavations, and deposited in the Museum at Naples.

Having witnessed the unsuccessful attempts of Dr. Sickler to unroll some of the Herculaneum manuscripts, it occurred to him that a chemical examination of their nature, and of the changes they had undergone, might suggest some method of separating the leaves from each other, and of rendering legible the characters impressed upon them. On communicating this opinion to Sir Thomas Tyrwhitt, he immediately placed at his disposal fragments which had been operated upon by Mr. Hayter and by Dr. Sickler: at the same time, Dr. Young presented him with some small pieces, which he himself had formerly attempted to unroll.

Davy was very soon convinced by the products of their distillation, that the nature of these manuscripts had been generally misunderstood; that they had not, as was usually supposed, been carbonized by the operation of fire, but were in a state analogous to peat, or to Bovey coal, the leaves being generally cemented into one mass by a peculiar substance which had formed, during the fermentation and chemical change of the vegetable matter composing them, in a long course of ages. The nature of this substance being once known, the destruction of it would become a subject of obvious chemical investigation.

It occurred to him, that as chlorine and iodine do not exert any action upon pure carbonaceous substances, while they possess a strong attraction for hydrogen, these bodies might probably be applied with success for the purpose of destroying the adhesive matter, without the possibility of injuring the letters of the Papyri, the ink of the ancients, as it is well known, being composed of charcoal. He accordingly exposed a fragment of a brown manuscript, in which the layers were strongly adherent, to an atmosphere of chlorine; there was an immediate action, the papyrus smoked, and became yellow, and the letters appeared much more distinct. After which, by the application of heat, the layers separated from each other, and fumes of muriatic acid were evolved. The vapour of iodine had a less distinct, but still a very sensible action. By the simple application of heat to a fragment in a close vessel filled with carbonic acid, or with the vapour of ether, so regulated as to raise the temperature very gradually, and as gradually to reduce it, there was a marked improvement in the texture of the papyrus, and its leaves were more easily unrolled. In all these preliminary trials, however, he found that the success of the experiment absolutely depended upon the nicety with which the temperature was regulated.

Different papyri having exhibited different appearances, he concluded that the same process would not apply in all cases; but even a partial success he considered as a step gained, and it served to increase his anxiety to examine in detail the numerous specimens preserved in the Museum at Naples, as well as to visit the excavations that still remained open at Herculaneum.

Mr. Hamilton, to whom these views were communicated, with that ardour which belongs to his character, entered warmly into a plan which might enable Sir Humphry Davy to accomplish his objects; and on his representation of them, the Earl of Liverpool and Viscount Castlereagh placed at his disposal such funds as were requisite for paying the persons whom it was necessary to engage in the process.

At the same time, Sir Humphry Davy had the honour of an audience of his late Majesty, then Prince Regent; and on witnessing the results, his Royal Highness was pleased to express his approbation, and graciously condescended to patronize the undertaking. Exulting in the prospect of success, and sanguine as to the importance of its results to literature, Davy embarked at Dover for the Continent, in order to proceed to Naples, on the 26th of May 1818.

During his journey, he was engaged in making observations on the comparative temperature of air incumbent upon land and water, with a view to account for the formation of mists over the beds of rivers and lakes. The results of this enquiry were embodied in a memoir, which was read before the Royal Society on the 25th of February 1819, and published in the Philosophical Transactions of that year. This paper, while it records the course of his observations, informs us of the direction of his route to the southern shores of Italy.

On the 31st of May, while passing along the Rhine from Cologne to Coblentz, we find him examining the relative temperature of the air, and of the water of that river. On the 9th, 10th, and 11th of June, he was making similar observations on the Danube, during a voyage from Ratisbonne to Vienna. On the 11th of July, he was similarly engaged on the Raab, near Kermond in Hungary. In the end of August he was on the Save in Carniola; in the middle of September on the Ironzo in the Friul; in the end of that month, on the Po, near Ferrara; and in the beginning of October, repeatedly on the Tiber, and on the small lakes in the Campagna of Rome, extending and multiplying his observations upon the formation of mists: from the results of which he established the law, that the formation of mist, on a river or lake, never takes place, if the temperature of the water be lower than that of the atmosphere; not even though the latter should be even saturated with vapour.

Possessed of this fact, he was enabled to explain a phenomenon which all persons who have been accustomed to the observation of Nature must have frequently witnessed, although it had never yet been philosophically explained, nor even fully discussed, viz.—the formation of mists over the beds of rivers and lakes, in calm and clear weather, after sunset.

Sir Humphry Davy thinks that whoever has considered the phenomena in relation to the radiation and communication of heat and nature of vapour, since the publication of the researches of MM. Rumford, Leslie, Dalton, and Wells, can scarcely have failed to discover their true causes.

"As soon as the sun has disappeared from any part of the globe, the surface begins to lose heat by radiation, and in greater proportions as the sky is clearer; but the land and water are cooled by this operation in a very different manner: the impression of cooling on the land is limited to the surface, and very slowly transmitted to the interior; whereas in water above 45° Fah., as soon as the upper stratum is cooled, whether by radiation or evaporation, it sinks in the mass of fluid, and its place is supplied by warmer water from below, and till the temperature of the whole mass is reduced nearly to 40°, the surface cannot be the coolest part.* It follows, therefore, that wherever water exists in considerable masses, and has a temperature nearly equal to that of the land, or only a few degrees below it, and above 45° at sunset, its surface during the night, in calm and clear weather, will be warmer than that of the contiguous land; and the air above the land will necessarily be colder than that above the water; and when they both contain their due proportion of aqueous vapour, and the situation of the ground is such as to permit the cold air from the land to mix with the warmer air above the water, mist or fog will be the result; which will be so much the greater in quantity, as the land surrounding or inclosing the water is higher, the water deeper, and the temperature of the water, which will coincide with the quantity or strength of vapour in the air above it, greater."

It will be remembered, that the rivers Inn and Ilz flow into the Danube below Passau; a circumstance which afforded Davy an excellent opportunity of confirming, by observation and experi-

* Water, when cooled down to 40°, expands in volume, and thus becomes specifically lighter; and therefore at that temperature remains at the surface.

ment, the truth of his theory. On examining the temperature of these rivers, at six o'clock A. M. June 11, that of the Danube was found to be 62°, that of the Inn 56.5°, and that of the Ilz 56°: the temperature of the atmosphere on the banks, where their streams mixed, was 54°. The whole surface of the Danube was covered with a thick fog; on the Inn there was a slight mist; and on the Ilz barely a haziness, indicating the deposition of a very small quantity of water. About one hundred yards below the conflux of the rivers, the temperature of the central part of the Danube was 59°; and here the quantity of mist was less than on the bed of the Danube before the junction; but about half a mile below, the warmer water had again found its place at the surface, and the mist was as copious as before the union of the three rivers.

After mists have been formed above rivers and lakes, Davy considers that their increase may not only depend upon the constant operation of the cause which originally produced them, but likewise upon the radiation of heat from the superficial particles of water composing the mist, which produces a descending current of cold air in the very body of the mist, while the warm water continually sends up vapour. It is to these circumstances, he says, that the phenomena must be ascribed of mists from a river or lake sometimes arising considerably above the surrounding hills. He informs us that he had frequently witnessed such an appearance during the month of October, after very still and very clear nights, in the Campagna of Rome above the Tiber, and on Monte Albano, over the lakes existing in the ancient craters of this extinguished volcano; and in one instance, on the 17th of October, before sunrise, there not being a breath of wind, a dense white cloud, of a pyramidal form, was seen on the site of Alban Lake, and rising far above the highest peak of the mountain. Its form gradually changed after sunrise; its apex first disappeared, and its body, as it were, melted away in the sunbeams.

Great dryness of the air, or a current of dry air passing across a river, he found, as we might have expected, to prevent the formation of mist even when the temperature of the water was much higher than that of the atmosphere.

Thus did our philosopher, during the course of his journey to Naples, by a series of observations and experiments, investigate a phenomenon connected with the deposition of water from the atmosphere, and which is not without an effect in the economy of nature; for verdure and fertility, in hot climates, generally follow the courses of rivers, and by the operation of the law he established, they are extended to the hills, and even to the plains surrounding their banks.

On his arrival at Naples, Sir H. Davy found that a letter from his Royal Highness the Prince Regent to the King, and a communication made from the Secretary of State for Foreign Affairs to the Neapolitan Government, had prepared the way for his enquiries, and procured for him every possible facility in the pursuit of his objects.

The different rolls of papyri presented very various appearances. They were of all shades, from a light chestnut brown to a deep black; some externally were of a glossy black, like jet, which the superintendents called "varnished;" several contained the umbilicus, or rolling-stick, in the middle, converted into dense charcoal. In their texture, also, they were as various as in their colours.

The persons to whom the care of these MSS. are confided, or who have worked upon them, have always attributed these different appearances to the action of fire, more or less intense, according to the proximity of the lava, which has been imagined to have covered the part of the city in which they were found; but the different conclusion at which Davy had arrived, from a chemical examination in England, was confirmed by a visit to the excavations that still remained open at Herculaneum.

These excavations are in a loose *tufa*, composed of sand, volcanic ashes, stones, and dust, cemented by the operation of water, which, at the time of its action, was probably in a boiling state. The theatre, and the buildings in the neighbourhood, are incased in this *tufa*, and, from the manner in which it is deposited in the galleries of the houses, there can be little doubt that it was the result of torrents laden with sand and volcanic matter, and descending, at the same time, with showers of ashes and stone still more copious than those that covered Pompeii. The excavation in the house in which the MSS. were found, had been filled up; but a building, which was said by the guides to be this house, and which, as is evident from the engraved plan, must at least have been close to it, at once convinced Davy that the parts nearest the surface, and, *à fortiori*, those more remote from it, had never been exposed to any considerable degree of heat. He found a small fragment of the ceiling of one of the rooms, containing lines of gold leaf and vermilion, in an unaltered state, which never could have happened had they been acted upon by any temperature sufficiently great to convert vegetable matter into charcoal.

The different states of the MSS. exactly coincide with this view, and furnish evidence of their having undergone a gradual process of decomposition. The loose chestnut papyri, he observes, were probably never wetted, but merely changed by the reaction of their elements, assisted by the operation of a small quantity of air; the black ones, which easily unroll, may be supposed to have remained in a moist state, without any percolation of water; while it is likely that the dense ones, containing earthy matter, have been acted on by warm water, which not only carried into the folds earthy matter suspended in it, but likewise dissolved the starch and gluten used in preparing the papyrus and the glue of the ink, and distributed them through the substance of the MSS.

As many of the papyri appear to have been strongly compressed when moist, in different positions, he thinks it probable that they had been placed on shelves of wood, which were broken down when the roofs of the houses yielded to the superincumbent mass. That the operation of fire is not at all necessary for producing such an imperfect carbonization of vegetable matter as that displayed by the MSS., is at once proved by an inspection of the houses at Pompeii, which was covered by a shower of ashes that must have been cold, as they fell at the distance of seven or eight miles from the crater of Vesuvius; and yet the wood of its buildings is uniformly found converted into charcoal, while the colours on the walls, most of which would have been destroyed or altered by heat, are perfectly fresh. Where papyri have been found in these houses, they have appeared in the form of white ashes, as of burnt paper, an effect produced by the slow action of the air penetrating through the loose ashes, and which has been impeded or prevented in Herculaneum by the *tufa*, which, as it were, hermetically sealed up the town, and prevented any decay, except such as occurs in the spontaneous decomposition of vegetable substances exposed to the limited operation of water and air—for instance, peat and Bovey coal.

Davy ascertained, that what the Neapolitans called varnish, was decomposed skin that had been used to infold some of the papyri, and which by chemical changes had produced a brilliant animal carbonaceous substance, which afforded by distillation a considerable quantity of ammonia, and left ashes containing much phosphate of lime.

Only one method, and that a simple and mechanical, though a highly ingenious one, had been adopted for unrolling the MSS. It was invented, in the middle of the last century, by Padre Piaggi, a Roman, and consists in attaching a thin animal membrane, by a solution of glue, to the back of the MSS. and then carefully elevating the layers by silk threads, which are gradually moved by the revolution of wooden pegs. Davy, shortly after his arrival, desired that the process of unrolling might be continued in his presence; and in considering the method in its general application, it occurred to him that some expedient might be used to facilitate the separation of the layers. For this purpose, he proposed to mix the solution of glue with a sufficient quantity of alcohol to gelatinize it, in order that it might not penetrate through three or four layers, which it was liable to do, when the texture of the papyrus was loose or broken, and the glue employed was in a liquid state. He also suggested the application of warm air for drying the papyrus, in the operation of attaching the membrane. It is not my intention to follow the chemist through all the various processes which he instituted for accomplishing his object; they may, however, be found in his paper entitled "Some Observations and Experiments on the Papyri found in the Ruins of Herculaneum," which was read before the Royal Society on the 15th of March 1821, and published in the Transactions of that year.

It only remains to be stated that Davy was not successful; but though the process of unrolling hitherto applied may not have received any considerable improvement from his science, and though he may not have succeeded in rendering any of the manuscripts legible, the failure is not to be attributed to his want of zeal, or to his want of skill, but solely, as it is generally admitted, to the unfortunate condition of the papyri.

It will be readily supposed that a failure in an investigation, from which he had anticipated so much advantage, was not sustained by a person naturally quick and irritable, without some demonstrations of impatience and dissatisfaction.

It was probably under the influence of such feelings, that he composed the conclusion of his memoir. "During the two months that I was actively employed in experiments on the papyri at Naples, I had succeeded, with the assistance of six of the persons attached to the Museum, and whom I had engaged for the purpose, in partially unrolling twenty-three MSS., from which fragments of writing were obtained, and in examining about one hundred and twenty others, which afforded no hopes of success; and I should gladly have gone on with the undertaking, from the mere prospect of a possibility of discovering some better result, had not the labour, in itself difficult and unpleasant, been made more so, by the conduct of the persons at the head of this department in the Museum. At first, every disposition was shown to promote my researches; for the papyri remaining unrolled were considered by them as incapable of affording any thing legible by the former methods, or, to use their own word, *disperati;* and the efficacy and use of the new processes were fully allowed by the *Svolgatori,* or unrollers of the Museum; and I was some time permitted to choose and operate upon the specimens at my own pleasure. When, however, the Reverend Peter Elmsley, whose zeal for the promotion of ancient literature brought him to Naples for the purpose of assisting in the undertaking, began to examine the fragments unrolled, a jealousy with regard to his assistance was immediately manifested; and obstacles, which the kind interference of Sir William A'Court was not always capable of removing, were soon opposed to the progress of our enquiries; and these obstacles were so multiplied, and made so vexatious towards the end of February, that we conceived it would be both a waste of the public money and a compromise of our own characters to proceed."

While in Italy, Sir H. Davy visited the baths of Lucca, and examined the waters which have given to that place so much celebrity. The results of his analysis formed the subject of a paper, which was published in the Memoirs of the Royal Academy of Sciences at Naples, of which Society he was a member.

At the spot where the temperature of the water was the highest, that is, in what are termed the *Caldi*, or hot baths, a considerable quantity of a substance is ejected, which produces a deposit of a brownish-yellow colour. Having collected a quantity of this deposit, he ascertained it to consist of oxide of iron and silica, in the proportion of about four parts of the former to three of the latter; and although the iron, at the time of its deposition, proved to be a *peroxide*, he thinks it probable that it existed in the water in the state of *protoxide*. He also supposes, that the oxide of iron and the silica had been dissolved together in the water, and been deposited from it in combination. He conceives that the fact which he had some years before noticed, of the analogy between the base of silica, and that of boracic acid, together with those observed by Berzelius, furnish sufficient reasons for classing silica amongst the acids, and for rendering it probable, that the oxide of iron and silica undergo a real chemical combination in the warm water, and that they are separated from the latter in consequence of the reduction of its temperature, after it has issued from the mountain.

A small portion of oxide of iron, he observes, is found in the waters of Bath, in which case it is also accompanied by silica; and he believes that, in many other instances, the oxide of iron is dissolved in water through the same agency: he moreover regards such facts as throwing considerable light upon the manner in which ochre is generated.

Sir Humphry Davy returned to England in 1820; and, on the 19th of June, in the same year, his venerable friend Sir Joseph Banks, who, notwithstanding his increasing infirmities, had continued to discharge the duties of President of the Royal Society to the latest period of his life, expired at his villa at Spring Grove, at the advanced age of seventy-seven.

Discussions necessarily arose as to the appointment of a proper successor, when persons of high and even exalted rank were proposed as candidates; but the more influential members of the Society at once found, in their own Council-chamber, two philosophers, whom they considered equally entitled to the honour of the situation, and equally well calculated for the discharge of its duties—Sir Humphry Davy, and Dr. Wollaston; but the latter having signified his fixed determination to decline competition, gave the whole weight of his influence to the former; and, under that arrangement, he received from the Council the compliment of being placed in the chair, until the general election of officers at the ensuing anniversary.

As the period of election approached, a few Fellows of the Society attempted to raise a clamour in favour of some more aristocratic candidate. To this circumstance, Davy alludes in the following letter.

TO THOMAS POOLE, ESQ.

MY DEAR POOLE, Grosvenor Street, June 1820.

I REGRET very much that you could not join me at dinner this day. To-morrow and the following day I shall be occupied by pressing affairs; but I shall be at home to-morrow till half-past eleven, and be most happy to see you.

I am not very anxious to remove "mists," for I feel that the President's chair, after Sir Joseph, will be no light matter; and unless there is a strong feeling in the majority of the body that I am the most proper person, I shall not sacrifice my tranquillity for what cannot add to my reputation, though it may increase my power of being useful.

I feel it a duty that I owe to the Society to offer myself; but if they do not feel that they want me, (and the most active members, I believe, do) I shall not force myself upon them.

I am, my dear Poole, very sincerely yours,

H. DAVY.

On the day of election, (November 30, 1820,) there was a feeble expression in favour of Lord Colchester, who was abroad at the time, and had not even been made acquainted with the intention of his supporters. Davy was therefore elected by an immense majority of votes. He was conducted into the meeting-room by his two friends, Mr. Davies Gilbert and Mr. Hatchett, and, to the gratification of every lover of science, he ascended the chair of Newton.

The value which he himself attached to this triumph, may be seen in his answer to a letter of congratulation from his friend Mr. Poole.

TO THOMAS POOLE, ESQ.

Grosvenor Street, Dec. 10.

MY DEAR POOLE,

I AM much obliged to you for your congratulations. The contest to my election defeated itself, for there were only thirteen votes for Lord Colchester out of nearly one hundred and sixty; and, had it been known that the attempt would have been made, I should have had at least double the number. The overwhelming majority has, however, shown the good opinion of the Society, which I trust and feel has not been diminished by my conduct in the chair.

I have never needed any motive to attach me to science, which I have pursued with equal ardour under all circumstances, for its own sake, and for the sake of the public, uninfluenced by the fears of my friends, or the calumnies of my enemies. I glory in being in the chair of the Royal Society, because I think it ought to be a reward of scientific labours, and not an appendage to rank or fortune; and because it will enable me to be useful in a higher degree in promoting the cause of science. To this cause, however, I should have been always attached, even had I not been in such good humour with the public, as I have reason to be.

Dr. Wollaston, my only formidable opponent in the beginning of the business, behaved like a true philosopher and friend of science; and Mr. Gilbert gave me his warmest support.

I am sorry that I have said so much about myself, but your long letter called for something. I wish I could say anything satisfactory on the subject of Captain Parry and his officers.* I have every reason to believe Lord Melville will do all he can on the occasion; no recommendation will be wanting from the Royal Society that can be given; but the Admiralty is bound by certain general rules, and will not do more in this instance than they would do in the case of a brilliant combat; but these brave and scientific navigators will be rewarded by a more durable species of glory.

Lady Davy joins me in kind remembrances.

I am, my dear Poole, sincerely yours,

H. Davy.

It was a question anxiously discussed by the friends of Davy, how far his elevation to the chair of the Royal Society was calculated to advance the cause of science, or to increase the lustre of his own fame. It will be readily perceived that this is a question perplexed by various conflicting interests,

* Mr. Poole informs me, that he "had been anxious to interest him, as President of the Royal Society, in favour of those brave and scientific navigators, particularly Lieutenant, now Captain, Liddon, who commanded the Griper, in Captain Parry's first voyage."

for it not only involves considerations relating to the character of the person, but to that also of the constitution and objects of the Society over which he is called upon to preside.

It is still doubtful whether the Royal Society, in the present advanced state of science, can derive advantage from possessing in its President, a philosopher actively engaged in any one branch of experimental enquiry. Sir Humphry Davy, in his first address from the chair, took occasion to observe, that "in the early periods of the establishment, when apparatus was procured with difficulty, when the greatest philosophers were obliged to labour with their own hands to frame their instruments, it was found expedient to keep in the rooms of the Society a collection of all such machines as were likely to be useful in the progress of experimental knowledge; and curators and operators were employed, by whom many capital experiments were made under the eyes of the Society.* But since the improvement of the mechanical and chemical arts has afforded greater facilities as to the means of carrying on experimental research, the transactions of

* The Charter of the Royal Society states that it was established for the improvement of NATURAL science. This epithet "*natural*" was originally intended to imply a meaning of which very few persons, I believe, are aware. At the period of the establishment of the Society, the arts of witchcraft and divination were very extensively encouraged; and the word natural was therefore introduced, in contradistinction to *super*-natural.

Although Sir Walter Scott, in his Demonology, alludes to the influence of this Society in diminishing the reigning superstition, he does not appear to have been acquainted with the circumstance here alluded to.

the Fellows, recorded by the Society, have, with some few exceptions, been performed in their own laboratories, and at their own expense."

In deciding upon the qualifications necessary for a President, this altered state of the Society must not be overlooked; nor can it be concealed, that the great discoveries of modern science have been achieved without any direct assistance from the Royal Society. Davy would have discovered the laws of electro-chemistry, and applied them for the decomposition of the alkalies—and the genius of Dalton would, by his atomic doctrine, have "snatched the science from the chaos of indefinite combination, and have bound it in the chains of number," had the Society never existed. At the same time, it must be allowed that, although it may not have directly advanced the progress of science, it has materially assisted its cause, by perpetuating the spirit of philosophical enquiry, and the love of scientific glory—by keeping alive upon the altar the sacred flame that genius may have kindled.

In the present state of science, the Royal Society imparts an inspiring principle to its various branches, by affording a rallying point, a centre of communication, to the philosophers of all nations, to whom kindred pursuits may render personal intercourse beneficial; and it becomes the paramount duty of the chief of this great republic so to preside over its arrangements, as to foster and encourage such an alliance. To this end, he must promote feelings of mutual kindness and liberality; and as the friend and umpire to all parties, it is his office to settle disagreement, to soothe disappointment, to kindle hope, and to subdue the vehemence which "engenders strife," in order that rivalship shall not pass into hostility, nor emulation degenerate into envy. It is evident that the talents and qualifications necessary for the discharge of such duties are of the highest order, extensive in their range, and diversified in their character. To which, however harshly the word may grate upon the ear of the philosopher, WEALTH must be considered as an essential and indispensable condition.

It may be fairly asked, whether a philosopher actively engaged in the pursuit of any branch of science, is so well adapted for the performance of such varied duties, as the person who possesses a general acquaintance with every department, but is not exclusively devoted to the investigation of any one branch; for, however correct may be his decisions, or unbiassed his judgment, the conduct of the former will ever be open to the charge of partiality, and the bare existence of such a suspicion, though it may be wholly groundless, will carry with it a train of evils. It is not in human nature to believe that the looker-on, and he who plays the game, are alike indifferent to the cards.*

On the other hand, it may be urged with some force, that the Presidency of the Royal Society should be reserved as the fair reward of scientific

* I state this opinion with the greater confidence, from a conviction that it is not singular. On conversing lately upon the subject with a gentleman to whom the Royal Society is deeply indebted for the sound judgment and discretion he displayed on occasions greatly affecting its interests, he replied, "Sir, we require not an Achilles to fight our battles, but an Agamemnon to command the Greeks."

labours, and not as an appendage to rank or to wealth:—that in England, we may in vain search amongst the aristocracy for one who feels a dignified respect for the sciences, and who is willing to afford that time which the faithful discharge of its duties would require.

To assert that Davy retained his popularity, or to deny that he retired from the office under the frown of a considerable party, would be dishonest. I would willingly dismiss this part of his life without too nice an examination; but I am writing a history, not an eloge.

As a philosopher, his claims to admiration and respect were allowed in all their latitude; but when he sought for the homage due to patrician distinction, they were denied with indignation. How strange it is, that those whom Nature has placed above their fellow men by the god-like gift of genius, should seek from their inferiors those distinctions which are generally the rewards of fortune. When we learn that Congreve, in his interview with Voltaire, prided himself upon his fashion rather than upon his wit; that Byron was more vain of his heraldry than of his "Pilgrimage of Childe Harold;" that Racine pined into an atrophy, because the monarch passed him without a recognition in the anti-room of the palace, and that Davy sighed for patrician distinction in the chair of Newton, we can only lament the weakness from which the choicest spirits of our nature are not exempt. Will philosophers never feel, with Walpole, that "a genius transmits more honour by blood than he can receive?" Had the blood of forty generations of nobility flowed in the veins of Davy, would his name have commanded higher homage, or his discoveries have excited greater admiration? But great minds have ever had their points of weakness: an inordinate admiration of hereditary rank was the cardinal deformity of Davy's character; it was the centre from which all his defects radiated, and continually placed him in false positions; for the man who rests his claims upon doubtful or ill-defined pretensions, from a sense of his insecurity, naturally becomes jealous at every apparent inattention, and he is suspicious of the sincerity of that respect which he feels may be the fruit of usurpation. If with these circumstances we take into consideration the existence of a natural timidity of character, which he sought to conquer by efforts that betrayed him into awkwardness of manner, and combine with it an irritability of temperament which occasionally called up expressions of ill-humour, we at once possess a clue by which we may unravel the conduct of our philosopher, and the consequences it brought upon himself during his presidency of the Royal Society. Nor must we leave out of sight that inattention to certain forms which, amongst those who are incapable of penetrating beyond the surface of character, passes for the offensive carelessness of superiority. Davy, after the example of Sir Joseph Banks, opened his house on one evening of the week for the reception of the Fellows of the Royal Society, and of other persons who were actively engaged in any scientific pursuit; but the invitations to these *soirées* were so irregularly managed, that they frequently gave offence, where they were intended to convey a compliment.

Conflicting opinions, respecting the management of the Royal Institution, most unfortunately also arose, and the President of the Royal Society, presuming upon his former alliance with that establishment, and upon the high obligations conferred upon it by the splendid discoveries he had achieved within its walls, was encouraged to exercise an authority which provoked an angry dissatisfaction; — schisms arose, and the party-spirit thus kindled in Albemarle Street soon spread to Somerset House.—But let us turn to the brighter part of the picture. In the discharge of the more important duties of his office, the Society received the full benefit of his talents and his virtues. At its meetings, he was constant in attendance, and dignified in his conduct and deportment; in its councils, he was firm in his resolves, correct in his judgments, zealous in his plans,* and impartial in his decisions. It has been said that he unduly favoured the pursuits of chemistry, to the injury and depression of the other branches of science: this is not the fact, as a reference to the Philosophical Transactions will amply testify; and the awards of the Copley medals will moreover show, that he alike extended the animating influence of his patronage to every part of natural philosophy. I am authorised by Sir James South to state, that during

* It was well known to his friends that, had his health not declined, he would have carried into effect a reform which he had long contemplated, and by which the Royal Society would have become, at once, more dignified and more useful.

his negotiations with the Government, for the purpose of securing to the British Nation the unalienable use of his splendid instruments, by the erection of a permanent observatory, Sir Humphry Davy was indefatigable in his exertions to accomplish so important an object; and that on one occasion, in the midst of severe illness, he travelled at no inconsiderable risk to London, from the distant seat of his friend Mr. Knight, to advocate a cause so essential, in his judgment, to the interests of Astronomy.

In the Autumn of 1821, Davy visited his mother and relatives at Penzance; upon which occasion he received from the inhabitants of the town, and from the gentlemen resident in its neighbourhood, a flattering testimony of respect, which made a deep and lasting impression upon his heart.

At a General Meeting, summoned for the purpose of taking into consideration some mode by which his fellow-townsmen might express their sense of his transcendent talents, and of the lustre which his genius had cast upon the place of his nativity:—It was unanimously RESOLVED—

"That a public dinner be given to Sir Humphry Davy, and that the Mayor be desired to wait upon him forthwith, in order to communicate the Resolution, and respectfully to request that he would appoint the day, on which it would be agreeable to him to meet their wishes."

On the day appointed, a deputation of Gentlemen proceeded in their carriages to the house of his mother, for the purpose of conducting him to the hotel, where an appropriate entertainment had been provided for the occasion.

The following letter evinces the sincere satisfaction which this visit afforded him.

TO THOMAS POOLE, ESQ.

MY DEAR POOLE, Penzance, July 28, 1821.

AN uncontrollable necessity has brought me here. Close to the Land's-end I am enjoying the majestic in nature, and living over again the days of my infancy and early youth.

The living beings that act upon me are interesting subjects for contemplation. Civilization has not yet destroyed in their minds the semblance of the great Parent of good.

Nature has done much for the inhabitants of Mount's Bay, by presenting to their senses all things that can awaken in the mind the emotions of greatness and sublimity. She has placed them far from cities, and given them forms of visible and audible beauty.

I am now reviving old associations, and endeavouring to attach old feelings to a few simple objects. I am, &c. H. DAVY.

Although the letter which follows is without date, I am unwilling to withhold it.

TO THOMAS POOLE, ESQ.

MY DEAR POOLE,

I HAVE been for some weeks absent from London, and have only just received your letter. When I return in the winter, I shall be glad to see Mr. A.—

I regret that your niece is so much indisposed. Lady Davy has been obliged to change her climate in consequence of a long-continued cough, but I am happy in being able to say she is now quite well.

After the fatigues of a long season in London, I am now enjoying the Highland scenery and sports with a purer pleasure, and I find, after the Alps and Pyrenees, even the mountains of Scotland possessing some peculiar beauties. You ought to come and see this country, which you would enjoy, both as a lover of nature and of man. The one is grand and beautiful; the other, moral, active, and independent.

I am, my dear Poole, your obliged friend,

H. Davy.

The Philosophical Transactions, during the Presidency of Sir Humphry Davy, evince the alacrity with which he redeemed the pledge given to the Society in his address on taking the chair—

"And though your good opinion has, as it were, honoured me with a rank similar to that of General, I shall be always happy to act as a private soldier in the ranks of Science."

Many years before even the identity of lightning and electricity was suspected, it had been observed, on several occasions, that the magnetism of the compass needle was not only destroyed, which might have been attributed to heat, but that it was even reversed by lightning.*

* Davy observes, that there are many facts recorded in the Philosophical Transactions, which prove the magnetising powers of light-

In the progress of electrical discoveries, the similarity between electricity and magnetism had not escaped observation,* and some philosophers had even attempted to establish the existence of an identity or intimate relation between these two forces. The experiments of Ritter, however, alone appeared to offer any confirmation of the supposed analogy; but so obscure was his language, and so wild and hypothetical his views, that few, if any, of them were repeated either in France or England, and their results were for a long time wholly disregarded.

In a work, entitled "Recherches sur l'identité des Forces Chimiques et Electriques," published by M. Oersted in the year 1807, the subject was resumed, and the author advanced the hypothesis,† which

ning: one in particular, where a stroke of lightning passing through a box of knives, rendered most of them powerful magnets.—*Philosophical Transactions*, No. 157, p. 520, and No. 437, p. 57.

* The phenomena of many crystallized minerals which become electric by heat, and develope opposite electric poles at their two extremities, offered an analogy so striking to the polarity of the magnet, that it seemed hardly possible to doubt a closer connection of the two powers. The developement of a similar polarity in the Voltaic pile pointed strongly to the same conclusion; and experiments had even been made with a view to ascertain whether a pile in a state of excitement might not manifest a disposition to place itself in the magnetic meridian.—*Herschel's Discourse*, p. 340.

† The hypothesis was this:—"In galvanism, the force is more *latent* than in electricity, and still more so in magnetism than in galvanism; it is therefore necessary to try whether electricity, in its *latent* state, will not affect the needle." This passage may be thus explained: When the Voltaic circuit is interrupted, it possesses opposite electrical poles; and when continuous, it no longer affects the electrometer, or the electricity becomes *latent*, which is the condition theoretically required for the manifestation of its mag-

twelve years afterwards conducted him to one of the most important discoveries of the present age, and which has given origin to a new science, termed ELECTRO-MAGNETISM.*

In the winter of 1819, Professor Oersted, Secretary to the Royal Society of Copenhagen, published an account of some experiments, in which the electric current, such as is supposed to pass from the positive to the negative pole of a Voltaic battery, along a wire which connects them, caused a magnetic needle near it to deviate from its natural position, and to assume a new one, the direction of which was observed to depend upon the relative position of the needle and the wire.†

It may be necessary to premise, that these experiments were conducted in a form which had never before suggested itself to the enquirer; *viz. with the two ends of the pile in communication with each other,*—a condition which enabled it to discharge itself freely: this circumstance will, at once, explain the reason of all preceding failures. It was never before suspected that the electric current, passing *uninterruptedly* through a wire, connecting the two ends of a Voltaic battery, was capable of

netic action: and the fundamental experiment of Oersted proved that, under these circumstances, the compass needle was affected.

* Mr. Herschel, in speaking of the pertinacity with which Oersted adhered to the idea of a necessary connection between electricity and magnetism, observes, that there is something in it which reminds us of the obstinate adherence of Columbus to his notion of the necessary existence of the New World.

† For this discovery, the President and Council of the Royal Society adjudged to M. Oersted the medal on Sir Godfrey Copley's Donation for the year 1820.

being manifested by any effect; the experiments, however, in question furnished an unequivocal test of its passage by its action on the magnetic needle; and which may be shortly stated as follows:

The opposite poles of a battery, in full action, were joined by a metallic wire, which, to avoid circumlocution, has been called the *uniting conductor*, or the *uniting wire*.

On placing the wire above the magnet and parallel to it, the pole next the negative end of the battery always moved westward, and when the wire was placed under the needle, the same pole went towards the east. If the wire was on the same horizontal plane with the needle, no declination whatever took place, but the magnet showed a disposition to move in a vertical direction; the pole next the negative side of the battery being depressed when the wire was to the west of it, and elevated when it was placed on the east side.

The extent of the declination occasioned by a battery, depends upon its power, and the distance of the uniting wire from the needle. If the apparatus is powerful, and the distance small, the declination will amount to an angle of forty-five degrees or more; but this deviation does not give an exact idea of the real effect which may be produced by galvanism; for the motion of the needle is counteracted by the magnetism of the Earth. When the influence of this latter power is destroyed by means of another magnet, the needle will place itself directly across the connecting wire: so that the real tendency of a magnet is to stand at right angles to an electric current. Such phenomena, being wholly at variance with the laws of simple electrical attraction and repulsion, are only to be explained upon the supposition that a new energy is generated by the action of the current of electricity thus brought into conflict, and which must be identical with, or nearly related to, magnetism.

It would also appear from the motions of the magnet, when differently placed with regard to the *uniting wire*, that this energy circulates, or performs a circular movement around the axis of the conductor, and thus drives the magnetic pole according to the direction of the needle with reference to such a current.

This important discovery was no sooner announced to the philosophical world, than Sir Humphry Davy, with his characteristic zeal, proceeded to repeat the experiments; and, with his usual sagacity, so to vary and extend them, as to throw new light upon this novel department of science. The facts he thus discovered, and the reasonings founded upon them, were communicated by him to the Royal Society in three successive memoirs.

THE FIRST, "On the Magnetic Phenomena produced by Electricity," was read on the 16th of November 1820.

THE SECOND, entitled "Farther Researches on the Magnetic Phenomena produced by Electricity; with some new Experiments on the properties of Electrified bodies, in their relations to conducting powers and Temperature," read July 5th, 1821.

THE THIRD, "On a new Phenomenon of Electromagnetism," read March 6th, 1823.

The principal experiments communicated in these

memoirs were performed with the battery belonging to the London Institution,* the once powerful apparatus at the Royal Institution having become old and feeble in his service.

The following letter contains an invitation to his friend Mr. Pepys, to witness his first experiment; a document so far valuable, as it fixes a date of some importance in the history of discovery.

TO WILLIAM HASLEDINE PEPYS, ESQ.

DEAR PEPYS, Grosvenor Street, Oct. 20, 1820.

THE experiment I wish to show you is no less than the conversion of electricity into magnetism; but it is a secret as yet.

I will come to you at twelve on Monday, in the Poultry. If you will be so good as to order the battery to be charged to-morrow, it will be ready for us on Monday.

Have you a dipping needle? This, and an air-pump, and the globe for taking sparks *in vacuo* by points of charcoal, are all we shall want.

Perhaps you will invite Dr. Babington, and our worthy friend Allen.

I will show you the opening of quite a new field of experiment. Ever yours very sincerely,

H. DAVY.

The discovery of Professor Oersted was limited to the action of the electric current on needles pre-

* I find from a note addressed to Mr. Pepys, that on the 21st of June, 1822, Davy worked the two batteries of 1000 plates each at the London Institution, before the Prince Royal of Denmark. The experiments were principally electro-magnetic.

viously magnetised. Davy ascertained that the *uniting conductor itself became magnetic*, during the passage of the electricity through it.* It was in consequence of having observed some anomaly, with respect to the way in which the uniting wire altered the direction of the magnet, that he was led to a conjecture which he immediately verified by a very simple experiment. He threw some iron filings on a paper, and brought them near the uniting wire, when immediately they were attracted by the wire, and adhered to it in considerable quantities, forming a mass round it ten or twelve times the thickness of the wire: on breaking the communication, they instantly fell off, proving that the magnetic effect entirely depended upon the passage of electricity through the wire.

Davy observes, it was easy to imagine that such magnetic effects could not be exhibited by the electrical wire, without its being capable of permanently communicating them to steel; and that, in order to ascertain whether such was the fact, he fastened several steel needles, in different directions, to the uniting wire, when those parallel to it were found to act like the wire itself, while each of those placed

* It would appear that M. Arago likewise discovered this fact at about the same period; but it is evident that the French and English philosophers arrived at the result independently of each other; for the experiments which led to it were made by Sir H. Davy in October 1820; while the September number of the "Annales de Physique," containing the first account of the Researches of M. Arago, was not received in London until the 24th of November in that year; and it may be farther observed, that the numbers of this journal were very commonly published several months after the affixed date.

across it acquired two poles. Such as were placed *under* the wire, the positive end of the battery being east, had north poles on the south of the wire, and south poles to the north. The needles *above* were in the opposite direction; and this was constantly the case, whatever might be the inclination of the needle to the wire. On breaking the connexion, the steel needles, placed *across* the uniting wire, retained their magnetism,* while those placed *parallel* to it lost it at the moment of disunion. The most extraordinary circumstances, however, connected with these experiments were, first, that *contact* with the uniting wire was not found necessary for the production of the effect,—indeed, it was even produced, though thick glass intervened; and, secondly, that a needle which had been placed in a transverse direction to the wire, merely for an in-

* M. Arago also, nearly at the same time, succeeded in communicating magnetism to the needle; but, at the suggestion of M. Ampère, it was effected in a different manner. A copper wire, by being rolled round a solid rod, was twisted into a spiral, so as to form a *helix*. It was easy, by passing the wire round the rod, in one direction or the other, to form a *dextrorsal* helix, proceeding from the right hand towards the left, as in the tendrils of many plants; or a *sinistrorsal*, or left helix, proceeding downwards from the left hand to the right above the axis. Into the cavity of a spiral thus formed, connecting the two poles of a battery, a steel needle wrapped in paper was introduced; and, in order to exclude all influence of the magnetism of the earth, the conchoidal part of the wire was kept constantly perpendicular to the magnetic meridian. In a few minutes the needle had acquired a sufficiently strong dose of magnetism; and the position of the north and south poles exactly agreed with M. Ampere's notion, that the electric current traverses the connecting wire in a direction from the zinc extremity of the pile to the copper extremity.

stant, was found as powerful a magnet as one that had been long in communication with it.

The distance to which magnetism is communicated by electricity, and the fact of its taking place equally through conductors and non-conductors, are circumstances which, in the opinion of Davy, are unfavourable to the idea of the identity of electricity and magnetism.

Davy subsequently ascertained by experiment, that the magnetic result was proportional to the quantity of electricity passing through a given space; and this fact led him to believe, that a wire electrified by the common machine would not occasion a sensible effect; and this he found to be the case, on placing very small needles across a fine wire connected with a prime conductor of a powerful machine and the earth. But as a momentary exposure in a powerful electrical circuit was sufficient to give permanent polarity to steel, it appeared equally obvious, that needles placed transversely to a wire at the time that the electricity of a common Leyden battery was discharged through it, ought to become magnetic; and this he found was actually the fact, and according to precisely the same laws as in the Voltaic circuit; the needle *under* the wire, the positive conductor being on the right hand, offering its north pole to the face of the operator, and the needle *above*, exhibiting the opposite polarity.

The facility with which experiments are made with the common Leyden battery, enabled him to ascertain various other important facts, respecting the communication of magnetism, which it would be inconsistent with the nature and limits of this work to particularize. I have merely offered a notice of the more prominent discoveries communicated by him in his first paper to the Royal Society, and which he concludes by observing, that "in consequence of the facts lately developed, a number of curious speculations cannot fail to present themselves to every philosophical mind; such as whether the magnetism of the earth may not be owing to its electricity, and the variation of the needle to the alterations in the electrical currents of the earth, in consequence of its motions, internal changes, or its relations to solar heat; and whether the luminous effects of the auroras at the poles are not shown, by these new facts, to depend on electricity. This is evident, that if strong electrical currents be supposed to follow the apparent course of the sun, the magnetism of the earth ought to be such as it is found to be."*

Davy never overlooked an occasion of applying theory to practice, and he therefore proposes, upon the principles developed in this paper, to make powerful magnets, by fixing bars of steel, or circular pieces of steel, fitted for making horse-shoe magnets, round the electrical conductors of buildings in elevated and exposed situations.

His second paper contains an account of experi-

* A very ingenious piece of apparatus was contrived to illustrate this theory by experiment; but I am uncertain as to whom the credit of it belongs. It consisted of a globe, containing metallic wires, arranged in relation to each other according to the electro-magnetic theory, when, by passing an electric current in the direction of the ecliptic, the poles became magnetic.

ments instituted with a view to gain some distinct knowledge on the subject of the relations of the different conductors to the magnetism produced by electricity. The results were decisive; but, without entering minutely into the theory of the subject which they so ably illustrated, these experiments cannot be clearly described, or successfully explained. The same observation will apply to the researches detailed in his third paper, announcing the discovery of a *new electro-magnetic phenomenon;* for, since they are inseparably connected with Mr. Faraday's beautiful experiments on *Magnetic Rotation,* I could scarcely expect to render my analysis of the memoir sufficiently intelligible, without entering at length upon that curious subject; I am unwilling, however, to refer the reader to the original paper in the Transactions, without offering a remark upon the *phenomenon,* which he says "is the *principal* object of the paper," but which we might conclude, from the hasty and imperfect manner in which he dismisses it, to have occupied a very subordinate place in his estimation. In his anxiety to examine and describe the rotations produced during this experiment, he bestows far too little attention upon the more, indeed I might say the *only,* important phenomenon of the cone of mercury which was elevated above each of the wires proceeding from the battery; and which, arising as it evidently did from a repulsive influence, clearly shows that the presence of electricity establishes between the particles of matter a repulsive energy, whether that matter be conducting, or non-conducting in its functions. This law, M. Ampère subsequently illustrated by a different form of experiment, and unfairly, as I must think, omitted even to notice Davy's prior result.

On the 20th of December 1821, Davy communicated to the Royal Society a memoir "On the Electrical Phenomena exhibited *in vacuo*."

It had been stated by Mr. Walsh, and the opinion had been subsequently supported by the researches of Mr. Morgan, that the electrical light was not producible in a perfect Torricellian vacuum; the latter gentleman also concluded that such a vacuum prevented the charging of coated glass.

An enquiry of greater importance can scarcely be imagined; involving in its train several of the most abstruse and difficult questions of corpuscular philosophy; as, whether electricity be a subtile fluid, or electrical effect the mere exhibition of the attractive powers of the particles of bodies; for, if it can be shown that these effects take place in a perfect vacuum, we shall advance towards the conclusion, that electrical phenomena depend upon the agency of an ethereal and transcendental fluid. It was under such an impression that Davy proceeded to determine, if possible, "the relations of electricity to space, as nearly void of matter as it can be made on the surface of the earth."

He was, in the first instance, led to suspect the accuracy of those conclusions at which Mr. Walsh and Mr. Morgan had arrived, from considering that, "in the most perfect vacuum which can be obtained in the Torricellian tube, vapour of mercury, though of extremely small density, must still always exist." I propose to follow our philosopher through the paths of this enquiry; and then, with all the deference due to such high authority, to state the objections which may be urged against his results.

First, then, as to the results he obtained with quicksilver in an apparatus simple, but well adapted at once to insure the most completely attainable vacuum, and to exhibit its capability of receiving a charge. In all cases where this vacuum was perfect, he found it to be permeable to electricity, and to be rendered luminous, either by the common spark, or by the shock from a Leyden jar; and, moreover, that the coated glass surrounding it became charged under such circumstances; but the intensity of the light in these experiments was always in proportion to the temperature, or, in other words, to the density of the mercurial vapour; and that at 20° below zero of Fahrenheit, it became so faint as to require considerable darkness to render it perceptible.

The great brilliancy, on the other hand, of the electrical light in pure, dense vapour of mercury, was beautifully displayed during the operation of boiling the metal in an exhausted tube. "In the formation and condensation of the globules of mercurial vapour, the electricity produced by the friction of the mercury against the glass, was discharged through the vapour with sparks so bright as to be visible in daylight."

The charge likewise communicated to the tinfoil was higher, the higher the temperature; at 0° Fahrenheit it was extremely feeble. This, like the phenomenon of the electric light, must, he thinks, depend upon the different density of the vapour of mercury.

But he was desirous of still farther refining his experiments, so as to exclude, as far as it was possible, the presence of any volatile matter; and in this part of the enquiry he displayed, in a very masterly manner, that happy talent in which he so far surpassed his contemporaries, of suggesting expedients and contriving new apparatus in order to vanquish practical difficulties.

To get rid of a portion of mercurial vapour, he employed a difficultly fusible amalgam of mercury and tin, which was made to crystallize by cooling in the tube; but, in this case, the results were precisely the same as when pure mercury had been used. He then attempted to make a vacuum above the fusible alloy of bismuth, but he found it so liable to oxidate and soil the tube, that he soon renounced farther attempts of this kind. Nothing discouraged, he determined to try the effects of a comparatively fixed metal in fusion. By melting freshly cut pieces of grain tin, in a tube made void after having been filled with hydrogen, and by long-continued heat and agitation, he obtained a column of fixed metal which appeared to be entirely free from gas; and yet the vacuum made above this exhibited the same phenomena as the mercurial vacuum, except that they were not perceptibly increased by heat: a fact which Davy must have anticipated, as he attributed the greater display of electrical light, at high temperatures, to the effect of increased density of vapour; it is therefore a matter of surprise that he did not give more importance to the phenomenon.

He made two experiments on electrical and magnetic repulsions and attractions in the mercurial vacuum, and he found that two balls, the one of platinum, the other of steel, properly arranged for the purpose, repelled each other, when the conducting wire to which they were attached was electrified in the most perfect mercurial vacuum, as they would have done in the usual cases: and that the steel globules were as obedient to the magnet as in the air; which last result, he observes, it was easy to have anticipated.

He also made some comparative experiments, with the view of ascertaining, whether below the freezing point of water the diminution of the temperature of the Torricellian vacuum diminished its power of transmitting electricity, or of being rendered luminous by it. To about twenty degrees, this appeared to be the case; but between twenty degrees above, and twenty degrees below zero, the lowest temperature he could produce by pounded ice and muriate of lime, it seemed stationary; and, as well as he could determine, the electrical phenomena were very nearly of the same intensity as those produced in the vacuum above tin.

"It is evident," he says, "from these general results, that the light (and probably the heat) generated in electrical discharges depends *principally* on some properties or substances belonging to the ponderable matter through which it passes: but they prove likewise that space, where there is no appreciable quantity of this matter, is still capable of exhibiting electric phenomena—viz. those of attraction and repulsion, &c.: a fact unquestionably favourable to the idea of the phenomena of electricity being produced by a highly subtile fluid or fluids, of which the particles are repulsive with respect to each other, and attractive of the particles of other matter.

However much we may admire the experimental address displayed in this paper, we must confess that its results are very far from being satisfactory. His having assumed, without proof, and even without examination, the theory that a perfect vacuum cannot be produced in the Torricellian tube, and made it the foundation of his reasonings, appears to me to have vitiated all his conclusions. Mr. Faraday has rendered it extremely probable, that a *limit* does actually exist to the production of vapour by bodies placed *in vacuo*,* beneath which they are perfectly fixed; and if this be true, it is evident that, at low temperatures, a perfect vacuum may be produced in the Torricellian tube; and it is highly probable that Davy did thus actually produce one in several of his experiments; especially in those where he found that, by a farther reduction of temperature, no farther diminution of electrical effect was perceptible: he had in fact arrived at this limit to vaporization, and therefore a farther reduction of temperature could not

* "On the Existence of a Limit to Vaporization. By M. Faraday, F.R.S. Corresponding Member of the Royal Academy of Sciences at Paris." Phil. Trans. 1826. See also a more recent paper by the same Philosopher in the first number of the new Journal of the Royal Institution.

possibly influence the phenomena. In this point of view, the electrical light would seem to be *primary*, or independent of foreign matter.—But though the premises be granted, let the reader pause before he hastens to any conclusion; for the cloud of mystery has not been dissipated, it has only changed its place. At the termination of his paper, Davy indulges in a conjecture subversive of every conclusion deduced from experiments *in vacuo*. "When the intense heat," says he, "produced by electricity, and the strong attractive powers of differently electrified surfaces, and the rapidity of the changes of state, are considered, it does not seem at all improbable, that the superficial particles of bodies, which, when detached by the repulsive power of heat, form vapour, may be likewise detached by electrical powers, and that they may produce luminous appearances in a vacuum free from all other matter, by the annihilation of their opposite electrical states."

During the course of the enquiry, Davy is led to suppose that air may exist in mercury, in the same invisible state as it does in water, that is, distributed through its pores; and that absorption of air may, therefore, explain the difference of the heights of the mercury in different barometers. This, it must be confessed, if true, is a most disheartening fact, as it at once precludes the possibility of any thing like accuracy in our barometers; but Mr. Daniell, to whom on all subjects of meteorology we are bound to pay the greatest deference, differs altogether from our philosopher upon this point, and he adduces a single observation which he thinks nearly disproves the supposition. "All fluids," says he,

" which are known to absorb air into their pores, invariably emit it when the pressure of the atmosphere is removed; but, upon an extensive examination of large bodies of mercury, variously heated in the vacuum of an air-pump, I never saw a bubble of air given off from the surface of the metal." Davy, it must be stated, obtained a far different result; but an observation of Mr. Daniell explains the cause of it. " Air," he continues, " will rise from the contact of the mercury with the glass in which it is contained, in exact inverse proportion to the care with which it has been filled, but it *never rises from the surface of the mercury alone.* The difficulty of properly filling a barometer tube, I attribute to the attraction between the glass and the air—not to that between the mercury and the air." *

On the 13th of June 1822, a memoir was read before the Royal Society,. " On the state of Water and Aëriform matter in cavities found in certain Crystals. By Sir Humphry Davy, Bart. P. R. S."

It is generally admitted by Geologists, that the greater number of the crystalline substances of the mineral kingdom must have been previously in a liquid state; but different schools have assumed different causes for their solution; some attributing the effect principally to the agency of water, others to that of heat.

In the paper under consideration, the author very freely avows himself as the champion of the latter doctrine.

* " Meteorological Essays and Observations," p. 363.—See also Bellani's experiments upon this subject, which are so satisfactory as to remove every doubt from the subject.

"When it is considered," says he, "that the solvent power of water depends upon its temperature, and its deposition of solid matters upon its change of state or of temperature, and that, being a gravitating substance, the same quantity must always belong to the globe, it becomes difficult to allow much weight to the arguments of the Wernerians, or Neptunists, who have generally neglected, in their speculations, the laws of chemical attraction.

"There are many circumstances, on the contrary, favourable to that part of the views of the Huttonians, or Plutonists, relating to the cause of crystallization; such as the form of the earth, that of an oblate spheroid flattened at the poles; the facility with which heat, being a radiating substance, may be lost and dissipated in free space; and the observations which seem to show the present existence of a high temperature in the interior of the globe."

He had often, he tells us, in the course of his chemical researches, looked for facts, or experiments, which might throw some light on this interesting subject, but without success, till it occurred to him, as he was considering the state of the fluid and aëriform matters which are found included in certain crystals, that these curious phenomena might be examined in a manner to afford some important arguments as to the formation of the crystal itself.

Having obtained, through the liberality of his friends, a variety of appropriate specimens of rock-crystal, he proceeded to submit them to experiment. Their cavities were opened by means of diamond drills, under either distilled water, oil, or mercury; the gas was then expelled from them by the introduction of slender wires, and the included fluids were drawn out by the aid of fine capillary tubes.

As soon as an opening was effected, the fluid under which the operation had been performed rushed into the cavity, and the globule of elastic fluid contracted so as to appear from six to ten times less than before the experiment. The fluid was found to be nearly pure water,—the gas appeared to be azote.

It was an interesting point to ascertain whether the same circumstances occurred in productions found in rocks which have been generally considered as of igneous origin, such as the basaltic rocks in the neighbourhood of Vicenza, the chalcedonies of which so often afford water. On examining such specimens, when, to obviate the possibility of any fallacy, they were previously ascertained to be impermeable to the atmosphere, analogous results were obtained: water, containing very minute quantities of saline impregnations, was found to be the fluid, and the gas, as in the former instances, was ascertained to be azote; but it was in a much more rarefied state than in the rock-crystals, being between sixty and seventy times as rare as atmospheric air.

The fact of azote being found in these cavities, he explains, by supposing that atmospheric air might have been originally included in the crystals, and that the oxygen had been separated from it by the attraction of the water; a conjecture which a direct experiment appeared to confirm.

In reasoning upon the vacuum, or rarefied state of the aëriform matter in the cavities of rock-crystals and chalcedonies, he very justly states, that the phenomenon cannot be easily accounted for, except on the supposition of their having been formed at a higher temperature than that now belonging to the surface of the globe: and he thinks it most probable that the water and the silica were in chemical union, and separated from each other by cooling, since there are strong grounds for believing that a liquid *hydrate of silica* would exist at high temperatures under pressure, and that, like all liquid bodies in the atmosphere, it would contain small quantities of atmospheric air. If this be granted, we may readily explain the phenomena presented by the gaseous and liquid matters in rock-crystal and chalcedony.

Thus then did Davy assail the Neptunists in their own camp, and vanquish them with their own weapons; for the fact, which had been confidently considered by the disciples of Werner, as, above all others, hostile to the idea of the igneous origin of crystalline rocks, namely, the existence of water in them, has been made to afford a decisive argument in favour of the very opinion it had been brought forward to oppose.*

In an appendix to the foregoing paper, the examination of two other crystals is detailed; the results afforded were very different from those of the preceding ones, but not less favourable to the theory of igneous origin. One of these crystals was found to contain a bituminous fluid; on piercing it under dis-

* I well remember with what triumph the late Dr. Clarke, in his popular lectures on Mineralogy at Cambridge, paraded a fine crystal containing water in its cavity. "Gentlemen," said he, "there is water enough in the very crystals in my cabinet to extinguish all the fires of the Plutonists."

tilled water, the water rushed in, and entirely filled the cavity, so that no aëriform matter but the vapour of the substance could have been present. The fact of almost a perfect vacuum existing in a cavity containing an expansible but difficultly volatile substance, must be considered as highly favourable to the theory of the igneous origin of crystals.

In the other crystal, the quantity of aëriform matter was unusually small in proportion to the quantity of fluid, and from the peculiarity of its motion, it appeared to be more likely to be compressed than rarefied elastic fluid; and in piercing the sides of the cavities, Davy found that this was the case; it enlarged in volume from ten to twelve times; the fluid was water, but the gas was too minute in quantity to be examined. There is but one mode of accounting for this phenomenon. The crystal must have been formed under an immense weight of atmosphere or fluid, sufficient to produce a compression much more than adequate to compensate for the expansive effects of heat.*

* An explanation which the experiments of Mr. Faraday, on the condensation of the gases, to be immediately described, will most fully justify.

CHAPTER XIII.

The Liquefaction of Chlorine Gas first effected by Mr. Faraday, and witnessed by the Author.—Sir H. Davy continues the investigation.—His paper on the application of Liquefiable Gases as mechanical agents.—Other probable uses of these bodies.—He proposes several methods to prevent the fumes which arise from Smelting-furnaces.—Importance of the subject. His Letters to Mr. Vivian.—The Government solicit the advice of the Royal Society on the subject of protecting the Copper Sheathing of Ships from the action of sea-water.—Sir H. Davy charges himself with this enquiry.—He proposes a plan of protection founded on Voltaic principles.—His numerous experiments.—He embarks on board the Comet steam-vessel bound to Heligoland, in order to try his plan on a vessel in motion.—He arrives at Mandal, lands, and fishes in the lakes.—The Protectors washed away.—He teaches the inhabitants of Christiansand to crimp fish.—He remains a few days at Arendal.—A Norwegian dinner.—The Protectors are examined and weighed.—Results of the experiment.—The steam-vessel proceeds up the Glommen.—He visits the great waterfall.—Passes into Sweden.—Has an interview with the Crown Prince of Denmark, and afterwards with Prince Christian at Copenhagen.—He visits Professor Oersted.—He proceeds to Bremen to see Dr. Olbers.—Returns to England.—His third paper read before the Royal Society.—Voltaic influence of patches of rust.—A small quantity of fluid sufficient to complete the circuit.—He receives from the Royal Society the Royal Medal.—The Progress of Voltaic discovery reviewed.—The principle is of extensive application.—The Author's researches into the cause of the solution of Lead in spring water.—

An account of the numerous trials of Protectors.—Failure of the plan.—Report of the French on the state of the protected frigate La Constance.—Dr. Revere's new plan of Protection.

EVERY incident, however trifling, if it relates to a great scientific discovery, merits the attention of the historian. As it accidentally occurred to me, and to me alone, to witness the original experiment by which Mr. Faraday first condensed chlorine gas into a liquid, I shall here state the circumstances under which its liquefaction was effected.

I had been invited to dine with Sir Humphry Davy, on Wednesday the 5th of March 1823, for the purpose of meeting the Reverend Uriah Tonkin, the heir of his early friend and benefactor of that name.* On quitting my house for that purpose, I perceived that I had time to spare, and I accordingly called in my way at the Royal Institution. Upon descending into the laboratory, I found Mr. Faraday engaged in experiments on chlorine and its hydrate in closed tubes. It appeared to me that the tube in which he was operating upon this substance contained some oily matter, and I rallied him upon the carelessness of employing soiled vessels. Mr. Faraday, upon inspecting the tube, acknowledged the justness of my remark, and

* Sir Humphry had expressed to me, on the preceding Thursday, at the Royal Society, his wish to purchase the old house in Penzance, which, as the reader will remember, was the early scene of his chemical operations; and, at his request, I conversed with Mr. Tonkin upon the subject; but it immediately appeared that the interest which the Corporation of Penzance possessed in the estate presented an insurmountable obstacle to the accomplishment of his object.

expressed his surprise at the circumstance. In consequence of which, he immediately proceeded to file off the sealed end; when, to our great astonishment, the contents suddenly exploded, and the oily matter vanished!

Mr. Faraday was completely at a loss to explain the occurrence, and proceeded to repeat the experiment with a view to its elucidation. I was unable, however, to remain and witness the result.

Upon mentioning the circumstance to Sir Humphry Davy after dinner, he appeared much surprised; and after a few moments of apparent abstraction, he said, "I shall enquire about this experiment to-morrow."

Early on the next morning, I received from Mr. Faraday the following laconic note:

DEAR SIR,

The *oil* you noticed yesterday turns out to be liquid chlorine. Yours faithfully,

M. FARADAY.

It is well known that, before the year 1810, the solid substance obtained by exposing chlorine, as usually procured, to a low temperature, was considered as the gas itself reduced into that form: Sir Humphry Davy, however, corrected this error, and first showed it to be a hydrate, the pure gas not being condensable even at a temperature of -40° Fahrenheit.

Mr. Faraday had taken advantage of the cold season to procure crystals of this hydrate, and was proceeding in its analysis,* when Sir Humphry Davy suggested to him the expediency of observing what would happen if it were heated in a close vessel; but this suggestion was made in consequence of the inspection of results already obtained by Mr. Faraday, and which must have led him to the experiment in question, had he never communicated with Sir Humphry Davy upon the subject. This avowal is honestly due to Mr. Faraday.

On exposing the hydrate, in a tube hermetically sealed, to a temperature of 100°, the substance fused, the tube became filled with a bright yellow atmosphere, and, on examination, was found to contain two fluid substances: the one, about three-fourths of the whole, was of a faint yellow colour, having very much the appearance of water; the remaining fourth was a heavy, bright yellow fluid, lying at the bottom of the former, without any apparent tendency to mix with it.

By operating on the hydrate in a bent tube hermetically sealed, Mr. Faraday found it easy, after decomposing it by a heat of 100°, to distil the yellow fluid to one end of the tube, and thus to separate it from the remaining portion. If the tube were now cut in the middle, the parts flew asunder, as if with an explosion, the whole of the yellow portion disappeared, and there was a powerful atmosphere of chlorine produced; the pale portion, on the contrary, remained, and when examined, proved to be a weak solution of chlorine in water, with a little muriatic acid, probably from the impu-

* The results are contained in a short paper in the Quarterly Journal of Science, vol. xv.

rity of the hydrate used. When that end of the tube in which the yellow fluid lay was broken under a jar of water, there was an immediate production of chlorine gas.

After several conjectures as to the nature of the changes thus produced, Mr. Faraday arrived at its true explanation; viz. that the chlorine had been entirely separated from the water by the heat, and condensed into a dry fluid by the mere pressure of its own abundant vapour. He subsequently confirmed these views by condensing chlorine in a long tube, by mechanical pressure, applied by means of a condensing syringe, and which farther enabled him to ascertain that the degree of pressure necessary for this effect was about that of four atmospheres.

To Mr. Faraday's paper upon this subject, published in the Philosophical Transactions for the year 1823, Sir Humphry Davy thought proper to add a "Note on the condensation of muriatic acid gas into the liquid form."

The circumstances under which this was effected are briefly these. On the morning (Thursday, March 6th,) after Mr. Faraday had condensed chlorine, Sir Humphry Davy had no sooner witnessed the result, than he called for a strong glass tube, and, having placed in it a quantity of muriate of ammonia and sulphuric acid, and then sealed the end, he caused them to act upon each other, and thus condensed the muriatic acid, which was evolved, into a liquid. The condensation of carbonic acid gas, nitrous oxide gas, and several others, were in succession treated with similar success; but, as I regard the discovery as strictly belonging to Mr.

Faraday, I shall confine myself to the relation of those experiments and deductions which, with equal justice, I must assign to Sir Humphry Davy.

He observes, "that the generation of elastic substances in close vessels, either with or without heat, offers much more powerful means of approximating their molecules than those dependent upon the application of cold, whether natural or artificial: for, as gases diminish only about $\frac{1}{450}$ in volume for every—degree of Fahrenheit's scale, beginning at ordinary temperatures, a very slight condensation only can be produced by the most powerful freezing mixtures, not half as much as would result from the application of a strong flame to one part of a glass tube, the other part being of ordinary temperature: and when attempts are made to condense gases into liquids by sudden mechanical compression, the heat, instantly generated, presents a formidable obstacle to the success of the experiment; whereas, in the compression resulting from their slow generation in close vessels, if the process be conducted with common precautions, there is no source of difficulty or danger; and it may be easily assisted by artificial cold in cases when gases approach near to that point of compression and temperature at which they become vapours."

On the 17th of April 1823, he communicated to the Royal Society a paper "On the application of Liquids formed by the condensation of Gases as mechanical agents."

He states that doubts may, for various philosophical reasons, exist as to the economical results to be obtained by employing the steam of water under great pressures, and at very elevated temperatures; but that no doubts can arise with respect to the use of such liquids as require for their existence even a compression equal to that of the weight of thirty or forty atmospheres; and where common temperatures, or slight elevations of them, are sufficient to produce an immense elastic force; and when the principal question to be discussed is, whether the effect of mechanical motion is to be most easily produced by an increase or diminution of heat by artificial means.

With the assistance of Mr. Faraday, he made several experiments on the differences between the increase of elastic force in gases under high and low pressures, by similar increments of temperature. In an experiment made with carbonic acid, its force was found to be nearly equal to that of air compressed to one-twentieth at 12° Fah. and of air compressed to one-thirty-sixth at 32 degrees, making an increase equal to the weight of thirteen atmospheres by an increase of twenty of temperature!

In applying, however, the condensed gases as mechanical agents, Davy admits that there will be some difficulty; "the materials of the apparatus must be as strong and as perfectly joined as those used by Mr. Perkins in his high-pressure steam-engine: but the small differences of temperature to produce an elastic force equal to the pressure of many atmospheres, will render the risk of explosion extremely small;" and he adds, "that if future experiments should realize the views here developed, the mere difference of temperature between sunshine and shade, and air and water, or the effects of evaporation from a moist surface, will be sufficient to produce results, which have hitherto been obtained only by a great expenditure of fuel."

If this be true, who can say that future generations shall not perform their voyages in *gas*-vessels, across the Atlantic Ocean, with no other fuel than that which a common taper may supply? I fear, however, that in this scientific reverie, Davy merely looked at the difference of the sensible temperatures, and entirely neglected, in his calculation, the quantity of heat rendered latent during the change of the liquid into the gaseous state; and which, perhaps, is far more considerable in the application of these fluids than in that of water; but even in this latter case, the great expenditure of heat in working the steam-engine, is in the portion rendered latent, and which cannot, by any contrivance, be brought again into operation, after it has performed its duty. That a philosopher who had, during the whole progress of his researches, directed such unremitting attention to the subject of Heat, should have wholly overlooked an objection arising out of one of its most familiar phenomena, is scarcely less extraordinary than his having, on another occasion,* advanced to a conclusion in

* I here allude to an anecdote related by Mr. Babbage, in his "Reflections on the Decline of Science in England;" a work, by the by, which strongly reminds me of a practical bull. A gentleman, anxious to escape the tax on armorial bearings, wrote a long letter to the Commissioners, stating I do not know how many reasons to show that he could never have used them; and, after all, sealed the letter with his own coat of arms! Had Mr. Babbage hoped to convince the reader that Science was actually on the

direct opposition to the very principle of Electricity, which his own discoveries had established.

Davy succeeded in liquefying gases by a method which, at first view, appears very paradoxical—*by the application of heat!* The method consists in placing them in one leg of a bent sealed tube, confined by mercury, and applying heat to ether, or alcohol, or water, in the other end. In this manner, by the pressure of the vapour of ether, he liquefied prussic gas and sulphurous acid gas; which gases, on being reproduced, occasioned cold.

There can be little doubt, he thinks, that these general facts of the condensation of the gases will

decline in this country, he should never have written a work which gives the lie to the title-page. Now for the anecdote.—"Meeting Dr. Wollaston one morning in the shop of a bookseller, I proposed this question: If two volumes of hydrogen and one of oxygen are mixed together in a vessel, and if by mechanical pressure they can be so condensed as to become of the same specific gravity of water, will the gases, under these circumstances, unite and form water? 'What do you think they will do?' said Dr. W. I replied, that I should rather expect they would unite. 'I see no reason to suppose it,' said he. I then enquired whether he thought the experiment worth making. He answered, that he did not, for that he should think it would certainly *not* succeed.

"A few days after, I proposed the same question to Sir Humphry Davy. He at once said, 'They will become water of course:' and on my enquiring whether he thought the experiment worth making, he observed that it was a good experiment, but one which it was hardly necessary to make, as it must succeed.

"These were off-hand answers, which it might perhaps be hardly fair to have recorded, had they been of persons of less eminent talent; and it adds to the curiosity of the circumstance to mention, that I believe Dr. Wollaston's reason for supposing no union would take place, arose from the nature of the electrical relations of the two gases remaining unchanged: an objection which did not weigh with the philosopher whose discoveries had given birth to it."

have many practical applications. They offer, for instance, easy methods of impregnating liquids with carbonic acid and other gases, without mechanical pressure. They afford means of producing great diminutions of temperature, by the rapidity with which large quantities of liquids may be rendered aëriform; and as compression occasions similar effects to cold, in preventing the formation of elastic substances, there is great reason to believe that it may be successfully employed for the preservation of animal and vegetable substances for the purposes of food.

Davy might also have added, that the same general views will explain natural and other phenomena not previously understood. They certainly afford a plausible explanation of the nature of *blowers* in coal-mines; and they may lead to more satisfactory views on other subjects of geology. They assign a limit to the expansive force of gas under increasing pressure, and account for effects connected with the *blasting* of rocks, which would otherwise appear anomalous.*

* In the year 1812, Mr. Babbage attempted to ascertain whether pressure would prevent decomposition: for this purpose, a hole about thirty inches deep, and two inches in diameter, was bored downward into a limestone rock, into which was then poured a quantity of strong muriatic acid, and a conical wooden plug, that had been previously soaked in tallow, was immediately driven hard into the mouth of the hole. It was expected either that the decomposition would be prevented, or that the gas developed would split the rock by its expansive force: but nothing happened. Now, it is most probable that a part of the carbonic acid had condensed into a liquid, and thus prevented that developement of power which Mr. Babbage had expected would have torn the rock asunder.

It may be stated, greatly to the honour of Davy, that there never occurred any question of scientific interest or difficulty in which he did not cheerfully offer his advice and assistance. Few Presidents of the Royal Society have ever exerted their influence and talents with so much unaffected zeal for the promotion of scientific objects, and for the welfare of scientific men. In the year 1821, the Great Hafod copper-works, in the neighbourhood of Swansea, were indicted for a nuisance, in consequence of the alleged destructive effects of the fumes which arose during the smelting of the ores. When we learn that the amount of wages paid by the proprietors of the works in this district exceeds 50,000*l.* per annum; that twelve thousand persons, at least, derive their support from the smelting establishments; that a sum of not less than 200,000*l.* sterling is annually circulated in Glamorganshire and the adjoining county, in consequence of their existence; that they pay to the collieries no less than from 100,000*l.* to 110,000*l.* per annum for coal; that one hundred and fifty vessels are employed in the conveyance of ore, and, supposing each upon an average to be manned by five seamen, that they give occupation to seven hundred and fifty mariners, a more serious calamity can scarcely be imagined than the stoppage of such works: we may therefore readily believe, that Davy entered most ardently into the consideration of some plan by which the fumes might be prevented, and the alleged nuisance abated.

Through the kind attention of my friend Mr. Vivian, I am enabled to insert the following letters.

TO JOHN HENRY VIVIAN, ESQ.

MY DEAR SIR, London, Jan. 9, 1822.

As you expressed a wish that I should commit to writing those opinions which I mentioned in conversation, when I had the pleasure of visiting you at Marino, after inspecting your furnaces and witnessing your experiments on the smoke arising from them, I lose no time in complying with your desire.

It is evident that the copper ore cannot be properly calcined without a copious admission of air into the furnaces, which must cause the sulphurous acid gas formed in the calcination to be mixed with very large quantities of other elastic fluids, which presents great mechanical, as well as chemical difficulties to its condensation or decomposition.

To persons acquainted with chemistry, a number of modes of effecting these objects are known. Of condensation, for instance, by water, by the formation of sulphuric acid, by alkaline lixivia, by alkaline earths, &c. Of decomposition, by hydrogen, by charcoal, by hydro-carbonous substances, and by metals; but to most of these methods there are serious and insurmountable objections, depending upon the diluted state of the acid gas, and the expenses required.

To form sulphuric acid, or to decompose by charcoal or hydrogen, or to condense by alkaline lixivia, or by alkaline earths, from the nature of the works, and of the operations for which they were intended, I conceive impracticable except at an expense that could not be borne; and the only processes which remain to be discussed are those by hydro-carbonous substances, and by the action of water.

There can be no doubt that the gas may be decomposed by the action of heated hydro-carbonous gases from the distillation of coal; but for this purpose there must be a new construction of the furnaces, and more than double, probably triple, the quantity of fuel would be required, supposing even the Swansea coal to contain the common average of bitumen; and this method must be infinitely more expensive, and liable to many more objections, than the one you have so ingeniously employed—absorption by water.

As water costs nothing, and as a supply is entirely in your power, the application of it offers comparatively few difficulties; and it has the great advantage of freeing the smoke from fluoric and arsenious compounds, which would not be perfectly effected by any other method.

The experiments of MM. Phillips and Faraday prove, that your shower baths have already entirely destroyed all the fluoric and arsenious fumes of the smoke, and by a *certain* quantity of water, the smoke may undoubtedly be entirely freed from sulphurous acid gas.

This, *your own* plan, is the one that I strongly recommend to you to proceed with, and, if necessary, to extend.

Perhaps you may find an additional shower bath near the colder part of the flue useful. I have no idea that steam passed into the hot part of the flue can be of the least service; but if passed out with the smoke through the stack, it may tend to convert such residual portion of sulphurous acid gas, exposed to fresh air, into sulphuric acid. Could you not likewise try a stream of *cold* water passing along the bottom of the horizontal flue?*

I do not think the advantages of your improvements can be fairly appreciated, till the effects of your smoke are determined by actual experiments and fair trials. Yours, &c.

H. DAVY.

TO THE SAME.

MY DEAR SIR, London, May 12, 1823.

I RETURN you my thanks for the copies you were so good as to send me of your work on the modes you have adopted for rendering copper smoke innoxious, &c. I have read it with very great pleasure, and I am sure there can be but one feeling, and that of strong admiration, at the exertions you have made, and the resources you have displayed, in subduing the principal evils of one of our most important national manufactures. I trust you will have no more trouble on this subject, and that it will only occur to you in an agreeable form, with the high approbation as well as grateful feelings of your neighbours; and that your example will be followed.

A Committee of the Royal Society has been formed for investigating the causes of the decay of copper sheeting in the Navy, as I mentioned to you. The Navy Board has sent us a number of specimens

* For the purpose of acting by its cooling power in condensing vapour, which would carry down sulphurous acid with it. It would likewise assist by direct absorption. H. D.

of copper in different stages of decay. We have our first meeting to examine them on Thursday, and I shall have much pleasure in communicating to you our results. I wish I could do it in person.

I am going into Hampshire on Sunday next to fish near Fordingbridge for a week, and to try the Avon and its tributary streams.

I was going to give you an account of some experiments which Mr. Faraday has made by my directions in generating gases in close vessels as liquids, but I find I have not time. I have already found an application of this discovery, which I hope will supersede *steam*, as a difference of a few degrees of temperature gives the elastic force of many atmospheres.

Hoping to see you soon, I am, with best respects to Mrs. Vivian, and love to the charming little Bessy,

My dear Sir, yours sincerely obliged,

H. Davy.

I proceed now to relate the history of an elaborate experimental enquiry, instituted for the purpose of ascertaining the chemical nature and causes of the well-known corrosive action of sea-water upon metallic copper; in order, if possible, to obviate that serious evil in naval economy—the rapid decay of the copper sheathing on the bottoms of our ships. An investigation which Sir Humphry Davy commenced in the year 1823, and prosecuted with his characteristic zeal and happy talent during a considerable period; when, at length, paradoxical as it may appear, the truth of his theory was completely established by the failure of his remedy !

From the several original documents which have been placed at my disposal, and from the valuable communications and kind assistance of my friend Mr. Knowles, I trust I shall be enabled to offer to the scientific reader a more complete and circumstantial history of this admirable enquiry than has been hitherto presented to the public.

The results he produced are equally interesting and important, whether we contemplate them biographically, as indicative of the peculiar genius by which they were obtained; or, scientifically, in their connexion with the electro-chemical theory, to the farther developement and illustration of which they have so powerfully contributed; or, economically, as the probable means by which the hand of Time may be averted, an increased durability imparted to rapidly perishable works of art, and monuments of human genius transmitted to posterity, in all their freshness, through a long succession of ages.

It is probable that, in the earliest period of naval architecture, some expedient* was practised, in order to protect ships' bottoms from the ravages of marine worms.† The use of metallic sheathing, how-

* Mr. Knowles, in his "Inquiry into the Means which have been taken to preserve the British Navy," observes, that the first sheathing was probably the hides of animals covered with pitch, or with asphaltum, which led to the use of thin boards, having, in some cases, lime, and in others lime and hair, between them and the bottom of the ships.

† The worms infesting the timber of ships are—the *Teredo*, the *Lepisma*, and the *Pholas*. The first of these, however, which was

ever, is of ancient date. The galley supposed to have belonged to the Emperor Trajan was sheathed with sheets of lead, which were fastened with copper nails.* The same metal was also used in the earlier periods of our naval history;† and it is worthy of remark, that the circumstances which led to its disuse, were the rapid corrosion of the *rother irons*, (from the formation of a Voltaic circle,) and the accumulation of sea-weed.

In the year 1761, copper plates were first used as sheathing on the Alarm frigate, of thirty-two guns;‡ a second underwent this operation in 1765, a third in 1770, four in 1776, nine in 1777; and, in the course of the three following years, the whole Bri-

imported from India, is by far the most destructive; and I am informed by Mr. Knowles, that it is more abundant at Plymouth than on any other part of the coast where there is a dock-yard; and although on the shores of England it is not of a very large size, yet it is a formidable enemy to the safety of those ships which have not a metallic sheathing to cover their bottoms. In the East Indies, and off the coast of Africa, the *Teredo* is of very large size; and holes have been bored by them in the timber of at least seven-eighths of an inch in diameter.

* Alberti Archeti.

† In the year 1670, an Act of Parliament was passed, granting unto Sir Philip Howard and Francis Watson, Esq. the sole use of the manufacture of milled lead for sheathing ships; and, in the year 1691, twenty ships had been sheathed with lead, manufactured by them, and which was fastened with copper nails.—See *Knowles's Inquiry*.

‡ The copper sheathing was removed from this ship in 1763, when all the iron was found to be much corroded, the pintles and braces nearly eaten through, and the false keel lost, from the decay of the keel staples and the bolt fastenings. Thus, in the very first coppered ship, the Voltaic effect, produced by the contact of copper and iron, was displayed in a very striking manner.

tish navy was coppered: an event which may be considered as forming an important era in the naval annals of the country.

The expense attending the use of copper for this purpose, in consequence of its corrosion and decay by salt-water, has always been felt as a serious objection to its use, and various suggestions have from time to time occurred, and numerous experiments been made, in the hope of obviating the evil,* but without any great degree of success.

The solution of the metal, however, has been found to vary in degree at different anchorages: at Sheerness, for instance, its rapidity is very great, in consequence of the copper being subjected to the alternate action of the sea, which flows in there from the British Channel, and to the flux of water down the two great rivers, the Thames and Medway, loaded, as they necessarily must be, with the products of animal and vegetable decomposition.

In order, if possible, to obtain a remedy for this evil, the naval departments of the Government requested, in the latter part of the year 1823, the advice of the President and Council of the Royal Society, as to the best mode of manufacturing copper sheets, or of preserving them, while in use, against the corrosive effects of oxidation.

Sir H. Davy charged himself with this enquiry;

* An experiment was tried by painting or varnishing their inner surfaces, but the use of brown paper which has been dipped in tar, and placed between the wood and copper, is now considered to be the best mode. A solution of caoutchouc spread on paper was tried on the bottom of Sir W. Curtis's yacht; but, on examination, it was pronounced to be less efficacious than tarred paper.

the results of which he communicated to the Royal Society, in three elaborate memoirs. The first was read on the 22nd of January 1824; the second, on the 17th of June, in the same year; and the third, and concluding paper, on the 9th of June 1825.

A very general belief prevailed, that sea-water had little or no action on *pure* copper, and that the rapid decay of that metal on certain ships was owing to its impurity. On submitting, however, various specimens of copper to the action of the sea-water, Sir H. Davy came to a conclusion, in direct opposition to such an opinion;* and Mr. Knowles informed me, in a late conversation upon the subject, that the attempts to purify the metal, since the Government has manufactured its own copper sheathing, has been the cause of its more rapid decay. It will however presently appear, that the relative durability of the metallic sheets must also be influenced by circumstances wholly independent

* In two instances, the copper (from the Batavier and from the Plymouth yacht) which had remained perfect for twenty-seven years, was found to be alloyed. In the former one there was an alloy of one three-hundredth part of zinc; and, in the latter, the same proportion of tin. On the other hand, in the case of the copper on the Tartar's bottom, which was nearly destroyed in four years, upon being submitted to chemical examination by Mr. Phillips, it was found to be very pure copper.

Alloys of copper have generally been found more durable than the unmixed metal; and various patents have been taken out for the fabrication of such compounds; but metallic sheets so composed have been found to be too hard and brittle, and not to admit of that flexibility which is necessary for their application to a curved surface; the consequence of which has been, that they have cracked upon the ship's bottom.

of their quality, some of which are very probably, even in our present advanced state of chemical knowledge, not thoroughly understood.

Sir H. Davy, on entering upon the examination of this subject, very justly considered, that to ascertain the exact nature of the chemical changes which take place in sea-water, by the agency of copper, ought to be the first step in the enquiry; for, unless the cause were thoroughly understood, how could the evil be remedied?

On keeping a polished piece of copper in contact with sea-water, the following were the effects which successively presented themselves. In the course of two or three hours, the surface of the metal exhibited a yellow tarnish, and the water in which it was immersed contracted a cloudiness, the hue of which was at first white, but gradually became green. In less than a day, a bluish-green precipitate appeared, and constantly continued to accumulate in the bottom of the vessel; at the same time, the surface of the copper corroded, appearing red in the water, and grass-green where it was in contact with air. Upon this grass-green matter carbonate of soda formed; and these changes continued until the water became much less saline. The green precipitate he ascertained to consist of an insoluble compound of copper, (which he thinks may be considered as a *hydrated sub-muriate*,) and hydrate of magnesia.*

According to his own views of the nature of chlorine, he immediately perceived that neither soda nor

* The Muriate of Magnesia is the most active salt in sea-water.

magnesia could appear in sea-water by the action of a metal, unless in consequence of an absorption or transfer of oxygen, which in this case must either be derived from the atmosphere, or from the decomposition of water: his experiments determined that the former was the source which supplied it. By reasoning upon these phenomena, and applying for their explanation his electro-chemical theory, which had shown that chemical attractions may be exalted, modified, or destroyed, by changes in the electrical states of bodies, he was led to the discovery of a remedy for the corrosion of copper, by the very principle which enabled him, sixteen years before, to decompose the fixed alkalies.

When he considered that copper is but weakly positive in the electro-chemical scale, and that it can only act upon sea-water when in a positive state, it immediately occurred to him that, if it could be rendered slightly negative, the corroding action of sea-water upon it would be null. But how was this to be effected? At first, he thought of using a Voltaic battery; but this could hardly be applicable in practice. He next thought of the contact of zinc, tin, or iron; but he was prevented for some time from trying this, by the recollection that the copper in the Voltaic battery, as well as the zinc, was dissolved by the action of dilute nitric acid; and by the fear, that too large a mass of oxidable metal would be required to produce decisive results. After reflecting, however, on the slow and weak action of sea-water on copper, and the small difference which must exist between their electrical powers; and knowing that a very feeble chemical action would be destroyed by a very feeble electrical force, he was encouraged to proceed; and the results were highly satisfactory and conclusive. A piece of zinc, not larger than a pea, or the point of a small iron nail, was found fully adequate to preserve forty or fifty square inches of copper,—and this, wherever it was placed, whether at the top, bottom, or in the middle of the sheet of copper, and whether the copper was straight or bent, or made into coils. And where the connexion between the different pieces of copper was completed by wires, or thin filaments of the fortieth or fiftieth of an inch in diameter, the effect was the same; every side, every surface, every particle of the copper, remained bright; whilst the iron, or the zinc, was slowly corroded.

A piece of thick sheet copper, containing on both sides about sixty square inches, was cut in such a manner as to form seven divisions, connected only by the smallest filaments that could be left, and a mass of zinc, of the fifth of an inch in diameter, was soldered to the upper division. The whole was plunged under sea-water; the copper remained perfectly polished. The same experiment was repeated with iron, and after the lapse of a month, the copper was in both instances found as bright as when it was first introduced; whilst similar pieces of copper, undefended, underwent in the same water very considerable corrosion, and produced a large quantity of green deposit in the bottom of the vessel.

Numerous other experiments were performed, and with results equally conclusive of the truth of the theory which had suggested them.

There was however one point which still remained for enquiry. As the ocean may be considered in its relation to the quantity of copper in a ship, as an infinitely extended conductor, it became necessary to ascertain whether that circumstance would influence the results. For this purpose, he placed two very fine copper wires, one undefended, the other defended by a particle of zinc, in a very large vessel of sea-water, which water might be considered to bear the same relation to so minute a portion of metal, as the sea to the metallic sheathing of a ship. The result was perfectly satisfactory. The defended copper underwent no change; the undefended tarnished, and deposited a green powder.*

Davy having thus satisfied his own mind as to the truth of his views, communicated to Government, in January 1824, the important fact of his having discovered a remedy for the evil of which

* During the course of some experiments in which I have been lately engaged, a simple mode of exhibiting the principle of protection occurred to me, which, I believe, has not before been suggested; at least, I cannot find any notice of such an experiment. As I consider it admirably calculated for illustration, I will here describe it. Let two slips of copper of equal size, the one protected with a piece of zinc, the other unprotected, be plunged into two wine-glasses filled with a solution of ammonia. In a short time, the liquor containing the unprotected copper will assume an intensely blue colour; the other will remain colourless for any length of time. The theory is obvious. When metallic copper is placed in contact with an ammoniacal solution, a protoxide of the metal is formed which is colourless,—and will remain so, if the contact of air be prevented; but on exposure to the atmosphere, it passes into a state of peroxide, which is dissolved by the ammonia, and produces an intensely blue solution. In the case of the protected copper, the metal is incapable of attracting a single atom of oxygen, in consequence of having been rendered negative by the zinc, and consequently no solution can take place.

they had complained; and that the corrosion of the copper sheathing of his Majesty's ships might be prevented by rendering the copper electro-positive, by means of the contact of tin, zinc, lead, iron, or any other easily oxidable metal; and that he was prepared to carry his plan into effect.

A proposition from a philosopher of such known science, and upon a subject of such great importance to the navigation and commerce of the country, immediately obtained all the attention it deserved; and an order was made that the plan of protection should, under the superintendence of Sir H. Davy, be forthwith tried upon the bottom of a sailing cutter.

To give to his discovery farther publicity, Sir Humphry requested that three models of ships might be exhibited in the spacious hall of the Navy Office in Somerset House; the copper of one of which he proposed should be protected by bands of zinc, that of another by plates of wrought iron soldered on the sheathing, while the third should have its copper exposed without any protection whatever.

These models were floated in sea-water for several months; and the experiment fully confirmed the results he had previously obtained in his laboratory. The models were from time to time examined by persons of the highest scientific character, as well as by others of great naval celebrity; and so alluring was the theory, and so conclusive the experiments, that, instead of waiting the result of the slow but more certain ordeal to which the plan had been submitted, it was immediately put into extensive practice, both in the Government service and on the bottoms of ships belonging to private individuals.

To those the least acquainted with the principles of Voltaic action, it was only necessary to state the proposition, in order to command their assent to its truth. The utility of the plan therefore was never questioned, but the claims of Davy to the originality of the invention were doomed to meet with immediate opposition.*

The correctness of the principle having been established, it became, in the next place, necessary to determine the most eligible metal to be used for protection; the proportion which it must bear to the surface of the copper-sheathing below the water-line; the form least likely to offer resistance to the sea, and to impede the sailing of the vessel; and lastly, its most convenient position on the ship's bottom. To ascertain these several points, Lord Melville and the Lords of the Admiralty desired the Commissioners of the Navy Board, and of the Dock-yards, to afford Sir Humphry every assistance and facility for prosecuting the necessary experiments; and he accordingly made many very exten-

* Amongst other counter-claims, there appeared, in a weekly publication entitled "The Mechanic's Magazine," a statement in favour of a person of the name of Wyatt, founded on the following advertisement in "The World" newspaper of April 16, 1791. "By the King's Patent, tinned copper sheets and pipes manufactured and sold by Charles Wyatt of Birmingham. These sheets, amongst other advantages, are particularly recommended for sheathing of ships, as they possess all the good properties of copper, with others obviously superior." It is unnecessary to observe that, except their object, there is nothing in common in the inventions of Davy and Wyatt. The superiority claimed by Wyatt consisted merely in coating the copper with some substance less corrosive by sea-water than that metal: an idea borrowed from the common practice of tinning copper vessels.

sive trials, not only on copper sheets which were immersed in the sea, but also on the bottoms of a considerable number of boats which had been coppered for that purpose, and exposed to the flow of the tide in Portsmouth harbour; upon which occasions he varied the nature as well as the proportions of the protecting metal. The results were communicated to the Royal Society, and they constituted the materials for his second memoir on the subject.

"When the metallic protector was from $\frac{1}{20}$ to $\frac{1}{110}$ parts of its surface, there was no corrosion nor decay of the copper; with smaller quantities, such as from $\frac{1}{200}$ to $\frac{1}{400}$, the copper underwent a loss of weight, which was greater in proportion as the protector was smaller; and, as a proof of the universality of the principle, it was found that even $\frac{1}{1000}$ part of cast iron saved a certain proportion of the copper.

"The sheeting of boats and ships, protected by the contact of zinc, or cast and malleable iron in different proportions, compared with those of similar boats and sides of ships unprotected, exhibited bright surfaces; whilst the unprotected copper underwent rapid corrosion, becoming first red, then green, and losing a part of its substance in scales. Fortunately, in the course of these experiments, it was proved that cast iron, the substance which is cheapest and most easily procured, is likewise most fitted for the protection of the copper. It lasts longer than malleable iron, or zinc; and the plumbaginous substance which is left by the action of sea-water upon it, retains the original form of the iron, and does not impede the electrical action of the remaining metal."

In the earlier stage of the investigation, it had

been suggested by Mr. Knowles, and several other persons, that by rendering the copper innoxious, it was probable sea-weeds might adhere to the sheets; but this objection he answered by stating, that negative electricity could not be supposed favourable to animal and vegetable life; and as it occasioned the deposition of magnesia, a substance exceedingly noxious to land vegetables, upon the copper surface, he entertained no difficulty upon that subject: in this, however, he was fatally mistaken. He found, after a trial of several weeks, that the metallic surface became coated with carbonate of lime and magnesia, and that, under such circumstances, weeds adhered to the coatings, and marine insects collected upon them; but at the same time he observed, that when the proportion of cast iron, or zinc, was below $\frac{1}{150}$, the electrical power of the copper being less negative, no such deposition occurred; and that although the surface had undergone a slight degree of solution, it remained perfectly clean: a fact which he considered of great importance, as it pointed out the *limits of protection;* and makes the application of a *very small* quantity of the oxidable metal more advantageous, in fact, than that of a larger one.

During the course of these experiments, many singular facts occurred to him, which tended to confirm his views of electro-chemical action. Amongst the various details which remained for his investigation, the relations between the surface of the protector, and that of the copper sheathing, under the different circumstances of temperature, saltness of the sea, and rapidity of the ship's motion, presented themselves as objects of great importance; and an opportunity occurred which enabled him to pursue them by actual observation and experiment.

In the month of June 1824, a steam-vessel, H.M. ship the Comet, was, at the express request of the King of Denmark, ordered to proceed to Heligoland, for the purpose of fixing with precision, by means of numerous chronometers, the longitude of that island, in order to connect the Danish with the British survey; and the Board of Longitude having recommended that the voyage should be extended as far as the Naze of Norway, for the purpose of ascertaining also the longitude of that important point, Sir H. Davy thought that this vessel would afford him the means of performing his desired experiments upon protected and unprotected copper sheets, when under the influence of rapid motion; and upon application to the Board of Admiralty, he obtained the entire disposal of the vessel after the required observations had been completed, as long as the season would allow her going to sea; and, that every facility might be afforded him, a skilful carpenter was put on board, to prepare whatever might be necessary for the prosecution of the enquiry.

For the following account of his adventures upon this occasion I am indebted to Dr. Tiarks, who, in his character of astronomical observer, superintended the expedition.

In the first instance, Davy directed to be constructed a number of oblong, rectangular, thin plates of copper, the surface of which should exceed that of a square foot: in the centre of these plates was fastened a slip of copper, by means of which other pieces of copper, which had small plates of iron of various dimensions attached to them, were fixed to the plate, by merely sliding them into the groove thus prepared for their reception. The plates were all carefully weighed previously to the experiment, and the pieces of iron were considered as representing the various proportions of iron and copper surfaces within whose limits Sir H. Davy had been led, by former experiments, to expect that the best proportion would be found. These plates were afterwards slipped into wooden frames, and nailed to the ship's side, over a piece of thick canvass, for the purpose of intercepting every possible communication between them and the copper sheathing.

It was proposed that, after each trip, these plates should be accurately weighed, in order to ascertain the loss which they severally might sustain from the corrosive action of the sea, while thus protected by different proportions of iron surface ; and, to ensure every possible accuracy, he carried with him the excellent balance, constructed by Ramsden, which is in possession of the Royal Society.

Sir H. Davy, accompanied by Lord Clifton, embarked at Greenwich on the 30th of June, and the vessel arrived at Heligoland on the 2nd of July. Here, as they remained not more than one day, the plates were not examined, although the Master expressed strong doubts as to their safety. The vessel then proceeded, by order of Sir Humphry, to Norway, a country which he was, for several reasons, very desirous of visiting, especially for the sake of determining a doubtful point in ornithology, upon which he subsequently corresponded with Professor Rheinhard, of Copenhagen.

The difference of longitude, also, between that country and Greenwich, not having been accurately ascertained, offered perhaps an additional reason for thus deviating from a course which, it must be confessed, was at variance with the original plan of the expedition.

After a severe gale of wind on the 4th of July, the vessel arrived, on the day following, at Rleve, near Mandal, and afterwards proceeded to this latter place, at which Davy remained for several days, during which interval the vessel made a tour to the Naze, and took in coal.

On the arrival of the vessel in the port, the plates were immediately examined; but, to the great disappointment of Sir Humphry, it was discovered that every one of the protectors had been washed away, and that most of the plates had sustained considerable injury.

With the country around Mandal he was much pleased; for, although it is far from being fertile, the scenery is rendered exceedingly striking and beautiful by the numerous lakes which wash the feet of high and sometimes perpendicular mountains, at that time clothed with the rich verdure of their summer herbage.

Sir Humphry made several excursions into the interior of the country, and derived much amusement from angling in the lakes; and had it not been from his own inspection of the roads, and the information which he collected respecting them, together with an indisposition of his fellow-traveller, Lord Clifton, he would have made an extensive land journey through the country; but, under the existing circumstances, he determined to return to

England through Denmark and Germany. He therefore at once resolved to take the steam-boat with him as far as Sweden, where the excellent roads would enable him, without inconvenience, to reach Gottenburg, and thence to continue his route through Denmark to Germany. The vessel proceeded accordingly to Christiansand, the chief town of a country of the same name.

Having been provided with some spare plates and protectors, he fixed them to the ship's side at Mandal, as he was informed that the voyage could be entirely performed within the rocks, with which the whole coast of Norway is so plentifully studded; but a short traverse through an open part of the sea, not far from Mandal, again defeated his object. The protectors were washed away, and no result was obtained.

At Christiansand he remained a few days, in order to try some new plates, which were constructed there under his own inspection. Upon this occasion he made an excursion to the falls of the Torjedahl, distant about six miles from the town. The river abounds with salmon, which were easily caught in their descent from the falls, by an apparatus contrived for that purpose. Sir Humphry amused himself by teaching the inhabitants the operation of *crimping*, and he declared the flavour of the fish to be superior to any salmon he had ever tasted.

It was at Christiansand that he became acquainted with the Norwegian race of ponies, so well adapted for mountainous countries; and which, at his recommendation, were afterwards introduced into England by Mr. Knight, of Downton Castle.

From Christiansand the vessel proceeded on her route eastward to Arendal, where she arrived on the 12th, after a passage of only a few hours. The route lay entirely within the rocks,—and so narrow were the passages, that the vessel could frequently not pass the rocks on either side without touching them.

At Arendal, which is the chief place of a remarkable mining district, Sir Humphry was well received by the Messrs. Dedehamys, two brothers, and the leading merchants of the place, with whom he made several excursions to the neighbouring mines. He was also invited by them to meet at their beautiful country seats the most respectable inhabitants of the town.

In the house of Mr. Dedehamy, Davy was introduced into Norwegian society, and, for the first time, had an opportunity of witnessing the customs and manners of the country.

A short time before dinner, the guests were summoned to partake of pickled fish, anchovies, and smoked salmon, with rum, brandy, and wine, which were placed on small tables in the drawing-room in which the company assembled. This custom of taking salt provisions, together with spirits, just before dinner, is very general in the North, and is considered as the best means of preparing the stomach, and of provoking an appetite for the approaching meal.

The very numerous party, which, with the exception of the hostess and her daughter, consisted entirely of men, were then ushered into two large rooms, one not being sufficiently spacious to accommodate them, and each person took his seat promiscuously. At the beginning of the dinner, large basins filled with sugar were carried round by the host's daughter, followed by a servant, from which each gentleman took a large handful. Sir Humphry, surprised at so singular a ceremony, enquired its meaning; when the host very good-humouredly answered, that in Norway they thought, if the wine was good it could not be spoiled by sugar,—and if bad, that it would be improved by it. Davy immediately followed the example of the company, and helped himself to the sugar.

Amongst the party present were several members of the Diet (Storthing), which had recently refused the applications of the King for various grants of money. This subject excited much animated conversation, and the majority of the persons present expressed their approbation at so bold and independent a measure. This called forth a political toast relating to the situation of their country; when the whole company, elated with wine and conversation, simultaneously burst forth into the national chorus of Norway, which had been composed as a prize poem during the short struggle against the union of that country with Sweden, and which was much admired by the Norwegians, and on all occasions sung by them with the utmost enthusiasm of feeling; but, notwithstanding the liberal politics of the party, they drank Sir Humphry's toast—"THE KING OF NORWAY AND SWEDEN"—with much apparent loyalty.

A succession of toasts followed, the last of which recommended "THE BRITISH CONSTITUTION AS A

MODEL FOR ALL THE WORLD." With this sentiment the festivities concluded—a momentary silence ensued; the custom of the country assigned to a stranger the honourable office of returning to the host and hostess the thanks of the company for their hospitable reception; all eyes were anxiously fixed upon the English philosopher; and as soon as he was made acquainted with the duty he was expected to perform, he rose from his seat, and in allusion to the sentiment so recently drunk in compliment to himself, he proposed as a concluding toast, "NORWEGIAN HOSPITALITY A MODEL FOR ALL THE WORLD."

From Arendal the vessel proceeded to Laurvig, where she stopped only a few hours; but Sir Humphry seized this opportunity to go on shore to view the country, and he afterwards weighed the copper plates which had been attached to the ship in Christiansand, as the vessel was now to cross that deep bay, at the bottom of which is situated Christiana, the capital of the kingdom. The few plates were found to be in good order; and the results, which however must be allowed to have been very incomplete, confirmed, as far as they went, the conclusions to which he had been led by former experiments, viz. that $\frac{1}{200}$ of iron surface was the proportion best calculated to defend the copper, without so over-protecting it as to favour the adhesion of marine productions; while they moreover proved that there is a mechanical as well as a chemical wear of the copper, which, in the most exposed part of the ship, and in the most rapid course, bears a relation to it of nearly 2 to 4.55.

The country increased in fertility towards the eastern parts of it; but it possessed much less beauty than the neighbourhood of Mandal.

As soon as Davy perceived that the vessel had to pass near the mouth of the Glommen, the largest river of Norway, he directed that she should enter it. Steam-boats appeared to have been entirely unknown in that part of the country. The inhabitants of the town of Frederickstadt were alarmed by the belief that the vessel was on fire, and they ran down to the beach in multitudes. As the vessel proceeded up the river, the people every where left their work, looked on awhile in silent amazement, and then shouted with delight.

The vessel anchored a mile below the great fall of the Glommen, called *Sarpen*, and which Davy visited on the following day (July 15). Three Kings of Denmark have visited this fall, and a name commemorates the spot whence they viewed this grand scene of nature. The fall is not one perpendicular descent, but consists of three sheets of water closely succeeding each other; and, by means of a barometer, he ascertained the entire altitude to be little more than a hundred feet. In comparing the character of this waterfall with those of the others he had visited, he observes, that size is merely comparative; and that he prefers the Velino at Terni, on account of the harmony that exists in all its parts. It displays all the force and power of the element, in its rapid and precipitous descent; and you feel that even man would be nothing in its waves, and would be dashed to pieces by its force. The whole scene is embraced at once by the eye, and the effect is almost as sublime as that of the Glommen, where the river is at least one hundred times as large; for the Glommen falls, as it were, from a whole valley upon a mountain of granite; and unless where you see the giant pines of Norway, fifty or sixty feet in height, carried down by it and swimming in its whirlpools like straws, you have no idea of its magnitude and power. Considering these waterfalls in all their relations, he is disposed to think, that while that of Velino is the most perfect and beautiful, the fall of the Glommen is the most awful.

On both sides of this fall are extensive saw-mills, with machinery of very imperfect construction. Davy spent some time with the proprietors of these mills, who were acquainted with the English language, and showed him every attention in their power. As an angler, he spoke with regret of the immense quantity of sawdust which floated in the water, and formed almost hills along the banks, and which, he observed, must be poisonous to the fish, by sometimes choking their gills, and interfering with their respiration.

From the Glommen the steam-vessel passed through the Svinesund to Strömstadt, the first town in Sweden beyond the frontier of Norway, from which Charles XII. essayed to besiege the neighbouring fortress of Frederickstadt in Norway. From Strömstadt, Davy set out on the 17th of July, and reached Gottenburg by land in two days, where he remained for a short time, in consequence of a slight indisposition. On his journey, he had a conversation with Oscar, the Crown Prince of Denmark, who, under the direction of Berzelius, had diligently devoted himself to the study of chemistry. He conversed with our philosopher upon various subjects connected with that science; and Davy, on his return to England, declared that he had never met with a more enlightened person.

The Crown Prince expressed great surprise, as indeed did every body in Sweden, on hearing that it was not Davy's intention to visit Professor Berzelius at Stockholm; and his astonishment was still farther increased, when he was informed by himself, that he came to Norway and Sweden with no other view than to enjoy the diversion of hunting and fishing! He however did by accident afterwards meet Berzelius, but his interview was but of short duration.

From Gottenburg he hastened to Copenhagen, where he renewed his acquaintance with Prince Christian of Denmark, cousin of the King, and heir presumptive of the crown; in whose company he had some years before observed an eruption of Mount Vesuvius. He also visited Professor Oersted, and earnestly requested that he might see the apparatus by which that philosopher had made those electro-magnetic experiments which had rendered his name so celebrated throughout Europe.

He next proceeded to Neuburg and Altona, where he intended to reimbark for England in the steam-vessel which had, during the interval of his continental tour, made a voyage to England, and was again on her way to the Elbe. At the suggestion, however, of Professor Schumacher, the astronomical professor at Copenhagen, but residing at Altona, in whose society he passed a great portion of his time, he accompanied that gentleman to Bremen, in order to make the acquaintance of the venerable Dr. Olbers, who, since his retirement from an extensive medical practice, had entirely devoted his time to the pursuit of his favourite science astronomy; as well as to be introduced to Professor Gauss, of Gottingen, who happened to be at that time carrying on his geodetical operations for the admeasurement of the kingdom of Hanover.

Davy expressed a great desire to see the telescope with which Dr. Olbers had discovered the two planets, Pallas and Vesta, and which to his great surprise turned out to be a very ordinary instrument. His personal intercourse with these two celebrated philosophers appeared to afford him the highest satisfaction; and he spent two days most agreeably in their society.

In his "Salmonia," he gives us some account of his adventures as an angler during this short excursion to Norway and Sweden. "All the Norwegian rivers," says he, "that I tried (and they were in the southern parts) contained salmon. I fished in the Glommen, one of the largest rivers in Europe; in the Mandals, which appeared to me the best fitted for taking salmon; and in the Arendal; but, though I saw salmon rise in these rivers, I never took a fish larger than a sea-trout; of these I always caught many—and even in the *fiords*, or small inland salt-water bays; but, I think, never any one more than a pound in weight. It is true that I was in Norway in the beginning of July, in exceedingly bright weather, and when there was no night; for even at twelve o'clock the sky was so bright, that I read the smallest print in the columns of a newspaper. I was in Sweden later—in August: I fished in the magnificent Gotha, below that grand fall, Trollhetta, which to see is worth a voyage from England; but I never raised there any fish worth taking. I caught, in this noble stream, a little trout about as long as my hand; and the only fish I got to eat at Trollhetta was bream."

He again embarked, on the 14th of August, on board the Comet steam-vessel, which had ascended the Weser as high as her draught of water would allow, and reached England, after a very boisterous passage, on the 17th of the same month; indeed, the vessel left the mouth of the Weser with a contrary wind, and the pilot was unwilling to put to sea, but Davy insisted on proceeding without delay. During the whole passage he suffered extremely from sea-sickness, and in a letter written to Professor Schumacher, shortly after landing, he remarks that "the sea is a glorious dominion, but a wretched habitation."

On the 9th of June 1825, Sir Humphry read before the Royal Society his third and most elaborate paper upon Copper sheathing, entitled "Farther Researches on the Preservation of Metals by Electro-chemical Means."

In this memoir, he states it to be his belief, that there is nothing in the poisonous nature of the copper to prevent the adhesion of weeds and testaceous animals; for he observes, that they will readily adhere to the poisonous salts of lead which commonly form upon the metal protecting the fore-part of the keel; and even upon copper, provided it be in such a state of chemical combination as to be insoluble. It is then, in his opinion, the *solution* of the metal—the *wear* of its surface, by keeping it smooth, which prevents the adhesion of foreign matter. Whenever the copper is unequally worn, deposits will, without doubt, rest in the rough parts, or depressions in the metal, and afford a soil or bed in which sea-weeds can fix their roots, and to which zoophytes and shell-fish can adhere; but there is another cause of foulness on the protected sheathing, arising from the deposit of earthy matter upon the copper, in consequence of its electro negative condition.

In relation to this subject, Davy has offered some observations upon the effects produced by partial formations of rust, which appear to me to be exceedingly interesting and important.

When copper has been applied to the bottom of a ship for a certain time, he says, a green coating, or rust, consisting of oxide, sub-muriate, and carbonate of copper, forms upon it; not equally throughout, but partially, and which, it is evident, must produce a *secondary*, partial, and unequal action, since those substances are negative with respect to metallic copper, and will consequently, by producing with it a Voltaic circuit, occasion a more rapid corrosion of those parts still exposed to sea-water: from this cause, sheets are often found perforated with holes in one part, after having been used for five or six years; while in other parts they are comparatively sound.* In like manner, the heads of the mixed

* The rusting of a common piece of iron, if carefully inspected, furnishes a beautiful illustration of this secondary action. The

metal nails, consisting of copper alloyed by a small quantity of tin, which are in common use in the Navy, give rise to oxides that are negative with respect to the copper, so that the latter is often worn into deep and irregular cavities in their vicinity.

A series of very interesting experiments, fully detailed in this memoir, which were instituted for the purpose of ascertaining the extent of the diminution of electrical action in instances of imperfect or irregular conducting surfaces, led him to the general conclusion, that a very small quantity of the imperfect or fluid conductor was sufficient to transmit the electrical power, or to complete the chain. This induced him to try whether copper, if nailed upon wood, and protected merely by zinc or iron on its *under* surface, or on that next the wood, might not be defended from corrosion: a question of great practical moment with regard to the arrangement of protectors. For this purpose, he covered a piece of wood with small sheets of copper, a nail of zinc of about $\frac{1}{100}$ part of the surface having been previously driven into the wood: the copper surface remained perfectly bright in sea-water for many weeks; and when the result was examined, it was found that the zinc had only suffered partial corrosion; that the wood was moist, and that, on the interior of the copper there was a considerable portion of revived zinc, so that the negative electricity, by its opera-

oxide, at first a mere speck, and formed perhaps by a globule of water, becomes negative with respect to the contiguous surface, and by thus forming a Voltaic circuit, exalts its oxidability, and the rust consequently extends in a circle.

tion, provided materials for its future and constant excitement. In several trials of the same kind, iron was used with similar results; and in all these experiments there appeared to be this peculiarity in the appearance of the copper, that unless the protecting metal below was in a large mass, there were no depositions of calcareous or magnesian earths upon the metal; it was clean and bright, but never coated. The copper in these experiments was nailed sometimes upon paper, sometimes upon the mere wood, and sometimes upon linen; and the communication was partially interrupted between the external and internal surfaces by cement; but even one side or junction of a sheet seemed to allow sufficient communication between the moisture on the under surface and the sea-water without, to produce the electrical effect of preservation. This last experiment of Davy is of greater importance than may at first appear, in showing what a small proportion of conducting fluid will complete a circuit, and in thus explaining phenomena, as I shall presently show, which might not otherwise be suspected to have an electrical origin.

These results upon perfect and imperfect conductors led him to another enquiry, important as it relates to the practical application of the principle, namely, as to the extent and nature of the contact or relation between the copper and the preserving metal. He was unable to produce any protecting action of zinc or iron upon copper through the thinnest stratum of air, or the finest leaf of mica, or of dry paper; but the action of the metals did not seem to be much impaired by the ordinary coating of oxide or rust; nor was it destroyed when the finest bibulous, or *silver-paper*, as it is commonly called, was between them, being moistened with sea-water. He made an experiment with different folds of this paper. Pieces of copper were covered with one, two, three, four, five, and six folds; and over them were placed pieces of zinc, which were fastened closely to them by thread; each piece of copper, thus protected, was exposed in a vessel of sea-water, so that the folds of paper were all moist.

It was found in the case in which a single leaf of paper was between the zinc and the copper, there was no corrosion of the copper; in the case in which there were two leaves, there was a very slight effect; with three, the corrosion was distinct; and it increased, till with six folds the protecting power appeared to be lost; and in the case of the single leaf, the result differed only from that produced by immediate contact, in there not being any deposition of earthy matter. Other experiments likewise proved that there was no absolute contact of the metals through the moist paper; for, although a thin plate of mica, as before stated, entirely destroyed the protecting effect of zinc, yet when a hole was made in it, so as to admit a very thin layer of moisture between the zinc and copper, the corrosion of the latter, though not prevented, was considerably diminished.

The experimental part of this paper concludes with an account of various trials to determine the electro-chemical powers of metals in menstrua out of the contact, or to a certain extent removed from the contact of air; in order, if possible, to diminish the rapid waste of the protecting metals. In the progress of these experiments he exhibits, in a most beautiful manner, the singular effect of different proportions of a fixed alkali, when mixed with sea-water, in rendering the iron, in its Voltaic connection with copper, more or less negative.

He terminates the paper with some observations of a practical nature, relative to the best modes of rendering iron applicable to the purposes of protection; but, as these have been already embodied in the investigation, it is not necessary to notice them farther in this place.

That I may give to the history of this subject all the perspicuity which it can derive from the connexion of its several parts, I shall now, in defiance of chronological order, proceed to consider his last Bakerian Lecture, "On the Relations of Electrical Changes," which was read before the Royal Society, on the 8th of June 1826. In which, after referring to his former papers on the chemical agencies of electricity, and the general laws of decomposition which were developed in them, he enters into some historical details respecting the origin and progress of electro-chemical science; being induced so to do, from a knowledge of the very erroneous statements which had been published upon the subject abroad, and repeated in this country. At the conclusion of this lecture, in reverting to the subject of Voltaic protection, he says: "A great variety of experiments, made in different parts of the world, have proved the full efficacy of the electro-chemical means of preserving metals, particularly the copper sheathing of ships; but a hope I had once indulged, that the peculiar electrical state would prevent the adhesion of weeds or insects, has not been realized; protected ships have often indeed returned, after long voyages, perfectly bright,* and cleaner than unprotected ships; yet this is not always the case; and though the *whole* of the copper may be preserved from chemical solution in steam-vessels (from the rapidity of their motion) by these means,—yet they must be adopted in common ships only so as to preserve a portion,—so applied, as to suffer a certain solution of the copper;† and an absolute remedy for adhesions is to be sought for by other more refined means of protection, and which appear to be indicated by these researches.

"The nails used in ships are an alloy of copper and tin, which I find to be slightly negative with respect to copper, and it is on these nails that the first adhesions uniformly take place: a slightly positive and slightly decomposable alloy would probably

* The Carnbrea Castle, a large vessel, of upwards of six hundred and fifty tons, was furnished with four protectors, two on the stern, and two on the bow, equal together to about 1-104th of the surface of copper. She had been protected more than twelve months, and had made the voyage to Calcutta and back. She came into the river perfectly bright; and, when examined in the dry-dock, was found entirely free from any adhesion, and offered a beautiful and almost polished surface; and there seemed to be no greater wear of copper than could be accounted for from mechanical causes.

† A common cause of adhesions of weeds or shell-fish, is the oxide of iron formed and deposited round the protectors. In the only experiment in which zinc has been employed for this purpose in actual service, the ship returned after two voyages to the West Indies, and one to Quebec, perfectly clean. The experiment was made by Mr. Lawrence, of Lombard Street, who states that the rudder, which was not protected, had corroded in the usual manner.

prevent this effect, and I have made some experiments favourable to the idea."

He next proceeds to state some circumstances, in addition to those he had formerly noticed, by which the electrical relations of copper are altered. "I found," says he, "copper hardened by hammering, *negative* to rolled copper;—copper (to use the technical language of manufacturers) both *over-poled* and *under-poled*,* containing, in one case, probably

* The *poling* of copper is an operation, the theory of which is involved in a great deal of mystery. Copper, when taken from the smelting furnace, is what is termed *dry*, that is, it is brittle, has an open grain and crystalline structure, and is of a purplish red colour. The following is the process by which it is refined, or *toughened*, by the process of *poling*. The surface of the melted metal in the furnace is, in the first place, covered with charcoal. A pole, commonly of birch, is then plunged into the liquid metal, which produces a considerable ebullition from the evolution of gaseous matter, and this operation is continued, fresh charcoal being occasionally added, so that the surface may be always kept covered, until the refiner judges from the assays that the metal is malleable. The delicacy of the operation consists in the difficulty of hitting the exact mark: if the surface should by accident be uncovered, it will return to its *dry* state; and should the process be carried too far, it will be *over-poled*, by which the metal would be rendered even more brittle than when in a *dry* state. When this is found to be the case, or, as they say, *gone too far*, the refiner directs the charcoal to be drawn off from the surface of the metal, and the copper to be exposed to the action of the air, by which means it is again brought back to its *proper pitch*, that is, become again malleable. Now the question is, what are the changes thus produced in the copper? Is the metal in its *dry* state combined with a minute portion of oxygen, of which *poling* deprives it, and thus renders it malleable? and does the *over-poling* impart to it a minute portion of carbon, and is copper, like iron, thus rendered brittle both by oxygen and carbon? Or, is the effect of the pole merely mechanical, that of closing the grain, and of altering the texture of

a little charcoal, and in the other a little oxide, *negative* to pure copper. A specimen of brittle copper, put into my hands by Mr. Vivian, but in which no impurity could be detected, was negative with respect to soft copper. In general, very minute quantities of the oxidable metals render the alloy positive, unless it becomes harder, in which case it is generally negative."

These are important facts, and should dispose those who may preside over judicial enquiries, to pause before they infer the inferiority of copper sheeting from the rapidity of its decay.*—I have

the metal? Something might be said in support of all these opinions. Mr. Faraday, who has attentively examined the subject, is unable to detect any chemical difference between *poled* and *unpoled* copper. On the other hand, when the metal is *over-poled*, it is found to oxidate more slowly, and its surface when in the furnace is so free from oxidation, that it is like a mirror, and reflects every brick in the roof. This certainly looks very much like carbonization —See "An Account of Smelting Copper, as conducted at the Hafod Copper-works; by J. H. Vivian, Esq."—*Annals of Philosophy*, vol. v. p. 113.

* This observation was suggested by an examination of a late judgment of the Court of Common Pleas, in the case of Jones *v.* Bright and others, on showing cause against rule for a new trial. This was an action brought by the Plaintiff against the Defendants for selling him copper, for the purpose of sheathing the ship Isabella, which, from the rapidity of its corrosion was inferred to have an inherent defect in its composition. In this case it was held, that with respect to *warranty*, there is a very wide difference as it applies to articles which are not the subject of manufacture, and those which are the produce of manufacture and of human industry. In the one case, it may be that no prudence, no care, could have guarded against a secret defect; in the other, by using due care, and providing proper materials, any defect in the manufacture may be guarded against. "In the case of the bowsprit, the man

now concluded a review of those admirable researches which led Sir Humphry Davy to suggest and mature a plan for arresting the corrosion of the copper sheathing of vessels by Voltaic action. Mr.

did not make the timber which composed the bowsprit; he merely cut it out, and fitted it to meet the purpose, and could therefore by no means have guarded against the rottenness in the centre of that bowsprit: but if a man makes copper, he may guard against inherent defects in that copper, by taking care that the copper contains a proper proportion of pure copper; and also by taking care that it is so well manufactured, that it does not drink in a greater quantity of oxygen than ought to be admitted into it, and that that oxygen, which of necessity gets in, (for some will,) shall be so distributed, that it shall not operate, as in the opinion of an intelligent witness the oxygen in this case did operate, by forming itself in patches, and thereby rendering it soft, and rendering the copper incapable of resisting the influence of salt-water— that he can guard against." With all due deference to the learned Judge, suppose it be shown that no human wisdom can guard against those circumstances by which a portion of the copper surface may be rendered more highly electro-positive, what becomes of the judgment? That the decay of copper sheathing is effected by extrinsic causes, and does not necessarily depend upon an inherent defect in the metal, may be proved in numerous ways. If it were owing to the quality of the copper, why should five, ten, or twenty sheets out of a hundred, made from the same charge of metal in a furnace and manufactured under precisely similar circumstances, be affected, and the remainder be perfectly sound? Why, again, should sheets, made from several distinct charges, placed on a particular vessel, be acted upon, while the same copper on other bottoms is not more than usually dissolved? Did any inherent defect exist in the metal, it surely must have equally affected the whole batch.

It is possible that, in some cases, in consequence of the sheets not having been properly cleansed before they are rolled, a portion of the oxide may be pressed into them by the rollers. In such a case, a Voltaic effect might be produced, and portions of the metallic surface rendered more electro-positive.

Babbage has said that he was authorised in stating, that "this was regarded by Laplace as the greatest of Davy's discoveries." I do not think, however, that it should be considered in the light of a separate performance: we do injustice to the philosopher by regarding it as an independent and isolated discovery; for it was the result of a long series of enquiries, which commenced by establishing the laws of electro-chemistry,—which led him to the decomposition of the alkalies and earths,—suggested to his unwearied genius a succession of novel researches, in a new field of enquiry,—and concluded, as we have seen, in producing the most striking results by means of the greatest simplicity. Not once during the progress of this enquiry had he any occasion to retrace his steps for the purpose of correction: justly has he observed in his last Bakerian Lecture, that, notwithstanding the various novel views which have been brought forward in this and other countries, and the great activity and extension of science, it is peculiarly satisfactory to find that he has nothing to alter in the fundamental theory laid down in his original communication; and which, after the lapse of twenty years, has continued, as it was in the beginning, the guide and foundation of all his researches.

The President and Council of the Royal Society appear to have been swayed by this consideration, when they adjudged to him "A Royal Medal,* for

* In the year 1825, His Majesty George IV communicated to the Royal Society, through Mr. Peel, his intention to found two gold medals, of the value of fifty guineas each, to be awarded annually by the Council of the Royal Society, in such a manner

his Bakerian Lecture on the relations of electrical changes, considered as the last link, in order of time, of the splendid chain of discoveries in chemical electricity, which have been continued for so many years of his valuable life."

Thus had Davy now received from the Royal Society all the honours they were capable of conferring upon him. In the year 1805, they adjudged to him the medal on Sir Godfrey Copley's donation for his various communications published in the Philosophical Transactions; in 1817, they awarded him the Rumford medals for his papers on combustion and flame; and in 1827, upon the grounds just stated, the President and Council expressed their unabated admiration by conferring upon him the only medal which remained for his acceptance —that which had been recently founded by their patron, his late Majesty.

Having thus disposed of the speculative part of his admirable enquiry, it will be interesting to pause in our narrative, in order to take a philosophical review of the progress of Voltaic discovery, in its relations to this particular object. It is a subject well calculated to afford a valuable lesson to the experimentalist, and at the same time to furnish illustrations, more striking even than that of the safety-lamp, of the necessity of that complicated species of machinery, without which the human mind is frequently unable to grapple with the simplicities of truth. It is true, that the fact of a galvanic effect

as shall, by the excitement of competition among men of science, seem best calculated to promote the object for which the Royal Society was instituted.

being excited by the contact of two dissimilar metals was noticed in the earliest stages of the enquiry, but it is equally evident that the phenomena which attended it, and the laws by which it was directed, required for their discovery and elucidation the assistance of the Voltaic battery. In reference to Davy, it may be here repeated, that the power of obtaining simple results, through complicated means, was one of the most distinguishing features of his genius.

It has been stated, that the Alarm frigate, the first coppered ship in our Navy, displayed very striking evidence of the effect of Voltaic action, in the rapid corrosion of its iron.* As early as 1783, after copper sheathing had become general, the Government issued orders that all the bolts under the line of fluitation should, in future, be of copper; but at that period, it was not possible that any idea could be entertained as to the true nature of the operation by which the iron was thus rapidly corroded, for it was only in the year 1797 that Dr. Ash noticed, for the first time, a phenomenon which was subsequently referred to the action of a simple Voltaic circuit.

It is a very curious circumstance in the history of

* Numerous are the instances of later date which might be adduced in illustration of the same fact; and it is now generally supposed that it may have been a frequent cause of ships foundering at sea. By oxidation, the volume of the iron at first increases, and then diminishes; in consequence of which the ship leaks, or, to use a technical expression, becomes "*bolt sick*." When the Salvador del Mundo was docked at Plymouth, in February 1815, the iron fastenings were in such a state of corrosion, that five planks near the bilge dropped into the dock when the water left her.

this subject, that, for many years after the Voltaic influence had been recognised as the agent in metallic corrosion, so far from the existence of the accompanying phenomenon of preservation being suspected, it was even supposed that the metals mutually corroded each other. At so late a date as 1813, we find Davy himself, in the letter addressed to M. Alavair, published in these memoirs,* dwelling upon the necessity of avoiding metallic contact, in order to prevent *corrosion*, without throwing out the most distant hint as to the simultaneous production of a converse effect.

The first distinct notice of a metal being preserved from oxidation by the contact of a dissimilar metal, is at once referred to a chemical law, without a reference even to its possible connection with Voltaic action; and, striking as the fact may now appear, it never attracted much attention. M. Proust observed, that although copper vessels be so imperfectly tinned, as to leave portions of the surface uncovered, still, in cooking utensils, we shall be equally protected from the poisonous effects of the former metal; because, says he, the superior readiness with which tin is oxidized and acted upon by acids, when compared with copper, will not allow this latter metal to appropriate to itself a single atom of oxygen.† The same chemist observes, that if lead be

* Page 13 of this Volume.

† As far as the principle of Voltaic protection goes, this may be very true; but it must be remembered that the acid generally present upon these occasions is acetic acid, which rises in distillation with water, so that at the boiling temperature it will be carried beyond the sphere of Voltaic influence, and may thus act upon the denuded copper as much as though tin were not present.

associated with tin, it will be incapable of furnishing to acids any saturnine impregnation, since the latter, being more oxidable than the former, will exclusively dissolve, and thus prevent the former from being attacked.

Whether the principle of Voltaic protection be applicable or not to the purpose of preserving copper sheathing, it is evident that it will suggest numerous other expedients of high importance in the arts, while it will explain phenomena previously unintelligible. By introducing a piece of zinc, or tin, into the iron boiler of the steam-engine, we may prevent the danger of explosion, which generally arises, especially where salt-water is used, as in those of steam-boats, from the wear of one part of the boiler. Another important application is in the prevention of the wear of the paddles, or wheels, which are rapidly dissolved by salt-water.

Mr. Pepys has also extended the principle for the preservation of steel instruments by guards of zinc: razors and lancets may be thus defended with perfect success. In the construction of monuments which are to transmit to posterity the record of important events, the artist will be careful in avoiding the contact of different metals: it is thus that the Etruscan inscriptions, engraved upon pure lead, are preserved to the present day; while medals of mixed metals of a much more recent date are corroded.

Numerous are the facts daily presented before us, which receive from this principle a satisfactory explanation. To the philosopher, the examination of its agencies will furnish a perpetual source of instruction and amusement; and I will here enumerate a few simple instances of its effects: in the first place, for the purpose of showing that, whenever a principle or discovery involves or unfolds a law of Nature, its applications are almost inexhaustible, and that, however abstracted it may appear, it is sooner or later employed for the common purposes of life; and in the next place, in the hope of convincing the reader, that there does not exist any source of pleasure so extensive and so permanent as that derived from the stores of philosophy. The saunterer stumbles over the stone that may cross his path, and vents only his vexation at the interruption; but to the philosopher there is not a body, animate or inanimate, with which he can come in contact, that does not yield its treasures at his approach, and contribute to extend the pleasures of his existence.

I well remember some years ago, that, in passing through Deptford, my curiosity was excited by the extraordinary brilliancy of a portion of the gilded sign of an inn in that town, while its other parts had entirely lost their metallic lustre. Having obtained a ladder, I ascended to the sign, in order, if possible, to solve the problem that had so greatly interested me: the mystery immediately vanished; for an iron nail appeared in the centre of the spot, which had protected the copper leaf for several inches around it. Any person may easily satisfy himself of the efficacy of such protection, in his rambles through the metropolis, by noticing the gilded, or rather coppered, sugar-loaves* so com-

* There is an excellent example at this time in the London Road leading to the Elephant-and-Castle.

monly suspended over the shops of grocers, when he will frequently perceive that the parts into which the iron supports have entered, unless the latter have been painted, shine preeminently brilliant. If a still more familiar example of the effect of a simple Voltaic circuit be required, it is afforded by the iron palisadoes, where the iron is constantly corroded at its point of contact with the lead by which it is cemented into the stone. These examples are not only interesting from their simplicity, but from their demonstrating the small quantity of a conducting fluid which is sufficient to transmit the electrical power, or to complete a simple circuit: a fact which, it will be remembered, the experiments of Davy had before established.*

As our knowledge advances, these principles will no doubt derive other illustrations, and be found capable of more extensive application; for as yet we are but in the infancy of the enquiry. I have lately been engaged in a series of experiments, the results of which, I confidently anticipate, will lead to some new facts connected with the changes produced on the negative metal of a Voltaic circuit; an account of which I hope shortly to submit to the Royal Society. I shall on this occasion merely notice one result, which appears to me to admit of an immediate application to one of the most important circumstances of life—the purity of water contained in leaden cisterns.

My attention has for several years been directed to the state of the water with which the metropolis is supplied; and upon having been lately requested

* Page 248 of this Volume.

to propose a remedy for preventing the action of a spring in the neighbourhood of London upon lead, which it had been found to corrode in a very rapid manner, I suggested the expediency of protecting the pipes and cisterns with surfaces of iron; but before such a plan was put in execution, I proposed to try its efficacy in the laboratory:—the first result was very startling; for, instead of preventing, as I had anticipated, I found that it greatly increased, the solution of the lead. After various experiments, I arrived at the conclusion, that lead, when rendered negative by iron, and placed in contact with weak saline solutions,—such, for instance, as common spring water,—was dissolved; in consequence of the decomposition of the salts and the transference of their elements according to the general law, the acid passing to the iron, and the alkali to the lead; and so powerfully is this latter body acted upon by an alkali, that, if a slip of it be immersed in a solution of potash or soda, its crystalline texture is so rapidly developed, that its surface exhibits an appearance similar to that presented by tin-plate, and which is designated by the term *moirée*.

I apprehend that most of the anomalous cases of the solution of lead in common water, which have for so many years embarrassed the chemist, may thus receive an explanation. An eminent physician lately informed me, that some time since he was called upon to attend a family who had evidently suffered from the effects of saturnine poison, and that he well remembers there was an iron pump in the cistern that supplied the water. Upon showing the results of my experiment to a no less eminent chemist, he was immediately reminded of a circumstance which occurred at Islington, where the water was found to corrode the lead in which it was received: in this vessel there was an iron bar; and the fact would not have attracted his notice, nor have been impressed upon his recollection, but from the unusual state of corrosion in which it appeared.

I shall conclude these observations by an account of "the change which some musket balls, taken out of Shrapnell's shells, had undergone," by Mr. Faraday, and which is published in the 16th volume of the Quarterly Journal of Science, for the year 1823. This history is not only interesting on account of the high chemical character of its author, but satisfactory as being in direct opposition to previously established facts; and cannot therefore have received any bias from preconceived theory.

"Mr. Marsh of Woolwich gave me some musket balls, which had been taken out of Shrapnell's shells. The shells had lain in the bottoms of ships, and probably had sea-water amongst them. When the bullets are put in, the aperture is merely closed by a common cork. These bullets were variously acted upon: some were affected only superficially, others more deeply, and some were entirely changed. The substance produced is hard and brittle; it splits on the ball, and presents an appearance like some hard varieties of hæmatite; its colour is brown, becoming, when heated, red; it fuses on platinum foil into a yellow flaky substance like litharge. Powdered and boiled in water, no muriatic acid or lead was found in solution. It dissolved in nitric acid without leaving any residuum, and the solution gave very faint indications only of muriatic acid. It is a *protoxide of lead*, perhaps formed, in some way, by the galvanic action of the iron shell and the leaden ball, assisted probably by the sea-water. It would be very interesting to know the state of the shells in which a change like this has taken place to any extent. *It might have been expected, that as long as any iron remained, the lead would have been preserved in the metallic state.*"

In one experiment, I found that a piece of lead protected by iron underwent solution in water containing nitrate of potash, while it resisted the action of very dilute nitric acid: upon this point, however, farther enquiry is necessary; for I subsequently failed in producing the same effect, owing, no doubt, to having employed too strong an acid.

Let us return from this digression to the subject of Sir Humphry Davy's Protectors. It only remains for me to relate the results which followed the practical application of the Voltaic principles which his various experiments had developed.

In the month of May 1824, directions were issued by the Lords of the Admiralty to protect, in future, the copper sheathing of all his Majesty's ships which might be taken into dock, upon the plan proposed by Sir Humphry Davy.

The protectors were bars of iron six inches wide at their base, three inches in thickness in their centre, and, in outward form, the segment of an extended circle. They were usually placed on each side of the ship in a horizontal position, viz. in midships about three feet under water—on the keel in a line with these—and the remainder in the fore and afterparts of the ship (about three feet under the line of fluitation), as far forward and abaft as the curvatures of their respective bodies would allow of their lying flat upon the surface of the copper.

As it is difficult by verbal description alone to convey a sufficiently distinct idea of this subject to persons unacquainted with naval architecture, I have introduced a sketch, exhibiting the *general* position of the *Protectors*, although they are necessarily exaggerated in size, or they would have appeared as mere specks upon the drawing.

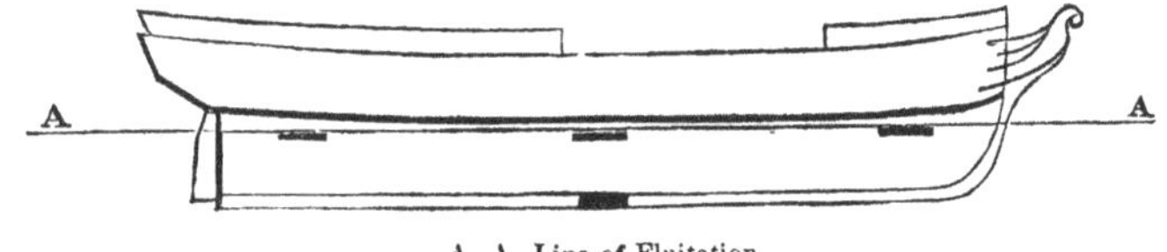

A. A. Line of Fluitation.

On several ships, some of the protectors, in the stem and the stern, were placed *vertically;* in which case they were fastened to the stems and stern-posts; and in this manner they were found to act more powerfully in preserving the copper, than when they were all placed horizontally. The ends of the protectors were rounded, in order to prevent any great resistance to the water, and they were fastened to the bottoms of the ships with copper bolts, the iron being counter-sunk to receive their heads, and the holes were then filled with carbonate of lime, or Parker's cement. To bring about the best possible contact of all the copper sheets, their edges, which lap over each other, where the nails are driven to fasten them to the ships, were rubbed bright, first with sand-paper, and finally with glass-paper.

Shortly after the ships thus protected were sent to sea, it was evident to all on board, from their dull sailing, that the bottoms had become very foul; and on being examined in dry docks, it was found that the copper was completely covered with sea-weed, shell-fish of various kinds, and myriads of small marine insects. Upon their removal, however, it was found, on weighing the sheets, that the copper had suffered little or no loss; thus proving that, although its practical application had failed from unforeseen circumstances, the principle of protection was true, and had fully justified the expectation of its success.

The copper near the protectors was much more foul than that at a greater distance from them; and there was, moreover, a considerable deposit of carbonate of lime, and of carbonate and hydrate of magnesia, in their vicinity

Sir Humphry Davy immediately suggested, as a remedy for this evil, that the bottoms should be scraped, and the copper washed with a small quantity of acidulous water; and he also proposed that the protectors should in future be placed under, instead of over, the copper sheathing. This plan was immediately adopted. Discs of cast-iron three and a half inches in diameter, and one-fourth of an inch in thickness, were let into the plank of the bottom of the Glasgow, of fifty guns, on the starboard side only—the larboard side having been left without any protection. These discs were in the proportion of one to every four sheets of copper, and over them were placed pieces of brown paper, and over the paper thin sheet-lead, so that the latter metal was in contact with the copper sheathing. A similar experiment was also tried on the Zebra, of eighteen guns, substituting, however, discs of zinc* for those of iron.

The bottom of the Glasgow was examined twelve months afterwards, when the discs of iron were found oxidated throughout, presenting in their appearance the characters of plumbago. The copper on the starboard side was preserved, but covered with weeds and shell-fish. The sheets on the larboard had undergone the usual waste, but were clean.

The Zebra was docked four years after the experiment had commenced, when the zinc protectors were perfect, and it did not appear that they had exerted any influence in preserving the copper, as it had wasted equally on both sides. It may be presumed in this case that the Voltaic circuit had by some fault in the arrangement been interrupted.

The apparent conversion of iron into a substance resembling plumbago, by the action of sea-water, has been frequently noticed. The protectors thus changed† were, to a considerable depth from the

* It would appear that Davy latterly preferred zinc to iron, as the protecting metal In a letter, dated October 1826, addressed to a ship-owner, who had made some enquiries of him upon the subject, he says—"The rust of iron, if a ship is becalmed, seems to promote the adhesion of weeds; I should therefore always prefer pieces of zinc, which may be very much smaller, and which, in the cases I have heard of their being used, have had the best effect."

† In the Annals of Philosophy (vol. v.) may be found a paper by Dr. Henry, on the conversion of cast iron pipes into plumbago. This change appears to have been effected by the action of water containing muriate of soda, and the muriates of lime and of magnesia.

surface, so soft as to be easily cut by a knife; but after being exposed for some time to the action of the atmosphere, they became harder, and even brittle. A portion of this soft substance having been wrapped in paper for the purpose of examination, and placed in the pocket of a shipwright, gave rise to a very curious and unexpected result: at first, the artist, like Futitorious with his chestnuts, thought he perceived a genial warmth; but the effect was shortly less equivocal; the substance became hot, and presently passed into a state of absolute ignition. Various theories have been suggested for its explanation: Mr. Daniell has advanced an opinion which supposes the formation of silicon, and thus accounts for the spontaneous ignition by the action of air.

The disadvantages which arise from the foulness of ships' bottoms, particularly when on foreign stations, where there are no dry docks to receive them, are so serious, that the Government was obliged, in July 1825, to order the discontinuance of the protectors on all sea-going ships; but directed that they should still be used upon all those that were laid up in our ports. When, however, an examination of the latter took place, they were found to

Cast iron contains a considerable portion of carbon; the change is therefore readily explained on the supposition of the removal of the principal metallic part by these salts. The muriates of lime and magnesia have been observed by Dr. Henry to discharge writing ink from the labels of bottles, to which they had been accidentally applied; and the same ingenious chemist has been baffled in his attempts to restore the legibility of ink upon paper which had been exposed to sea-water. The texture of the paper was not injured, but the iron basis of the ink, as well as the gallic acid, was entirely removed.

be much more foul than those which had been in motion at sea: shell-fish of various kinds had adhered to them so closely, that it was even necessary to use percussion to remove them, which not only indented the copper, but in many instances actually fractured it.

Under all these discouraging circumstances, the unwelcome conviction was forced upon the agents of Government, that the plan was incapable of successful application, and it was accordingly altogether abandoned in September 1828.

Such were the results of the experiments carried on in the ports of England, for the protection of copper sheathing, from the success of which Sir H. Davy justly expected honours, fame, and reward. That his disappointment was great, may be readily imagined, and it is supposed to have had a marked influence upon his future character. It is much to be regretted that his vexation should have been heightened by the unjust and bitter attacks made upon him by the periodical press, and by those subalterns in science, who, unable to appreciate the beauty of the principle he had so ably developed, saw only in its details an object for sarcasm, and in its failure an opportunity for censure; while those whose stations should have implied superior knowledge, in the pride and arrogance of assumed contempt, sought a refuge from the humiliation of ignorance.

That Davy was severely hurt by these attacks, is a fact well known to his friends. In a letter to Mr. Children he says: "A mind of much sensibility might be disgusted, and one might be induced to say, Why should I labour for public objects, merely to meet abuse?—I am irritated by them more than I ought to be; but I am getting wiser every day—recollecting Galileo, and the times when philosophers and public benefactors were burnt for their services." In another letter he alludes to the sycophancy of a chemical journal, which, after the grossest abuse, suddenly turned round, and disgusted him with its adulation. "I never shake hands," says he, "with chimney sweepers, even when in their May-day clothes, and when they call me '*Your honour*.'"

While the trials above related were proceeding in the ports of England, the naval department of France was prosecuting a similar enquiry; and as experiments of this nature are conducted with greater care, and examined with superior science, in that country, it may not be uninteresting to the English reader to receive a detail of the examination of the bottom of La Constance frigate, in which the protectors bore a much larger proportion to the copper surface than was ever practised in the British navy. This document, I may observe, is now published for the first time.

"The inspection of the bottom of the frigate La Constance, has given rise to some interesting observations on the effect of protectors, and it has confirmed the fact before advanced of the great inconvenience which attends the application of too large a proportion of the protecting metal.

"The surface of this metal, which was of cast iron, placed on each side of the keel, and in long scarphs of iron plates situated towards the stem and stern-post and the water line, appeared to have been about the 1-30th part of the surface of the copper, instead of the 1-250th part as now practised.

" The galvanic action has been extreme, both in rapidity and intensity. The scarphs are entirely destroyed, and have absolutely disappeared; and we should have been ignorant of their having ever existed, had we not been informed of the fact, and observed dark stains which marked their position, and discovered the nails still entire by which they had been fastened.

" The plates, which were in the first instance about three inches thick, were covered throughout their whole length by a thick, unequal coating, spotted with yellow oxide. This was principally owing to the absorption of about twenty-five per cent. of its weight of water. Under this, the iron was as soft as plumbago, and there remained scarcely an inch of metal of its original metallic hardness.

" The bulky and irregular appendage (the protectors) at the lower part of the ship's bottom caused a great noise in the sea, in consequence of the dead water which it occasioned, and doubtless lessened the speed of the vessel. But that which contributed most to this unfortunate result, was the exceedingly unclean state of the copper, arising from the excess of the iron employed: this, carried to so great an extent, having the effect of extracting matter from the water, which, forming a concretion on the sheets, enabled the marine animals the more easily to attach themselves. The sheathing was covered with a multitude of *lepas anatifera*, shells with five valves, suspended by a pedicle of three or four centimetres long, collected into groups; of *lepas tintinnabulum*, a shell with six valves; of oysters with *opercula;* of *polypi,* &c. No part of the bottom was free from them.

"Below, the copper was certainly preserved from oxidation; and up to within a few sheets of the water line, it did not appear to be worn. But to save expense, it was obliged to be cleansed without removal, by rubbing it hard with bricks and wet sand, which has succeeded very well in restoring its copper colour."

The following is the description of shells above enumerated:—

Genus *Anatifa,* Encyclopedia.—(*Lepas,* Linnæus.)

Fig. 1.

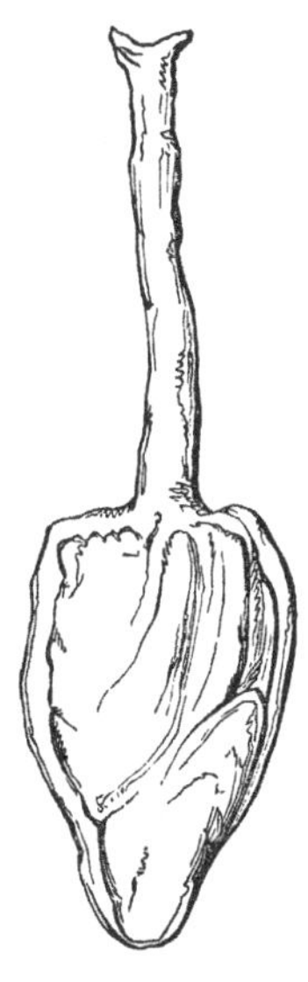

FIG. 1. Smooth *Anatifa* (*Lepas anatifera,* Linn.)—Shell consisting of five valves, of which two larger and two smaller ones are opposite to each other; and a fifth, which is narrow, is arched and rests upon the ends of the first four: these valves are not connected by any hinge; they are held together by the skin of the animal, which lines their interior and opens in front by a longitudinal separation. Their colour is orange during the life of the animal. The base of the shell is united to a fleshy tube, tendinous, cylindrical, susceptible of contraction, saffron-coloured, becoming brown and black in drying.

Fig. 2.

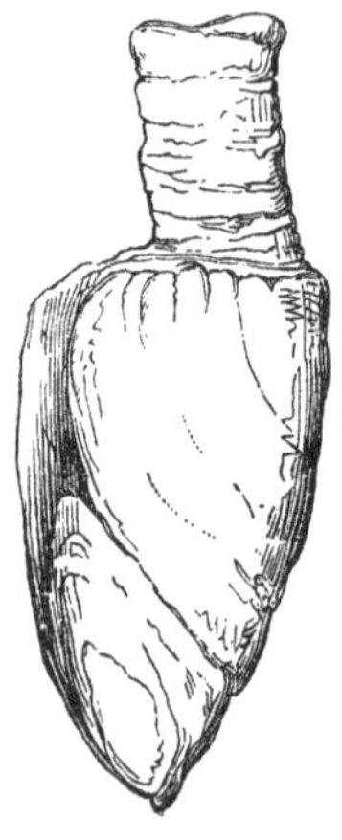

FIG. 2. Smooth *Anatifa*, as seen from the other side, the pedicle dry and contracted.

Fig. 3.

FIG. 3. Smooth *Anatifa*, as seen in front, showing the longitudinal separation.

Genus *Anomia*.

Shell with valves, unequal, irregular, having an operculum; adhering by its operculum; valve usually pierced, flattened, having a cavity in the upper part;

the other valve a little larger, concave, entire; operculum small, elliptical, bony, fixed on some foreign body, and to which the interior muscle of the animal is attached.

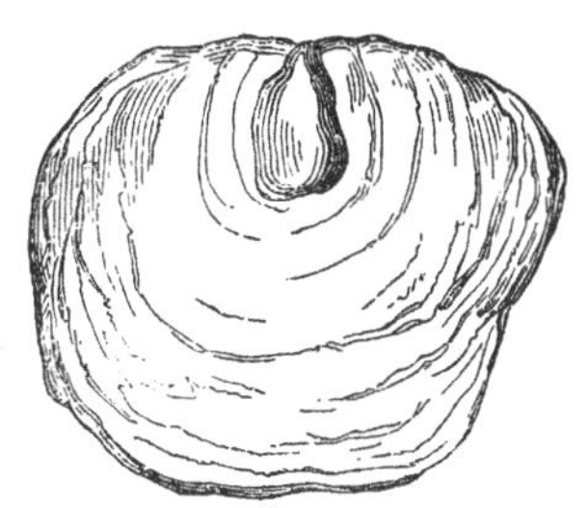

Species, Onion-peel *Anomia.*—(*Ephippium*, Linn.)

Shell common, whitish and yellowish, found in the Mediterranean and the ocean.

Besides the abovementioned species, which were found in large quantities, there were also some muscles and oysters.—(*Mytilus afer. Baccina.*—Linn. Gmel. 3358.)

Genus "*Balane de Blainville.*"
(*Balanite*, Encyclopedia.—*Lepas*, Linn.)

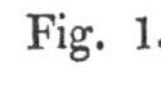

Fig. 1.

FIG. 1. Tulip Balanus.—(*Lepas tintinnabulum*, Linn.)

Shell with six unequal valves articulated by a scaly suture, of which the edges appear to be finely crenellated in the cavity; the form of the valves is conical, aperture ample, and nearly quadrangular.

Operculum composed of four triangular pieces crenellated and marked with very projecting transverse striæ, which appear to extend from the top to the bottom; the two posterior pieces are perpendicular, and are applied to the hinder partition of the cavity of the shell; they are terminated by two conical prolongations, of which the points are sharp and diverging. The two foremost pieces are placed in the aperture, in an oblique direction. The colour of this balanus from clear red to violet and brown.

Fig. 2.

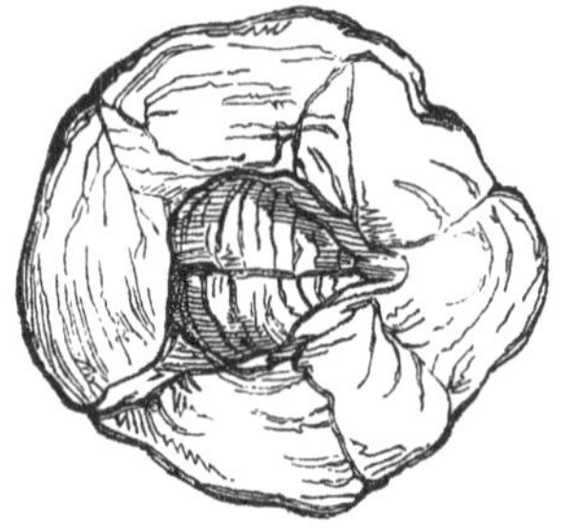

FIG. 2. View of the upper part of the Tulip Balanus.

Fig. 3.

FIG. 3. View of the base.

Genus Oyster, (*Ostrea.*)

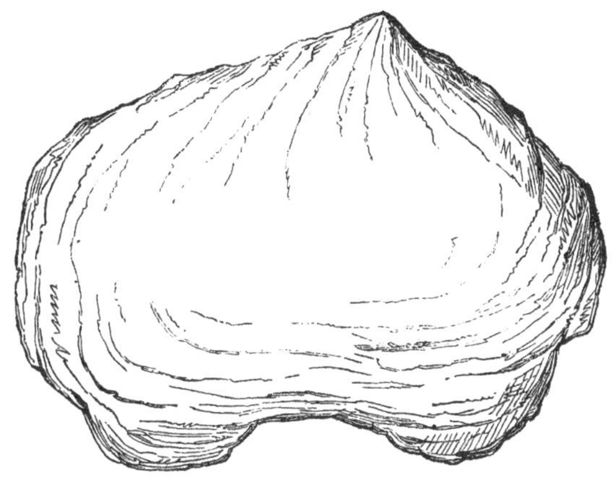

Species of oyster, nearly similar to the common oyster, (*Ostrea edulis,*) and of the Huître cuilier, (*Ostrea cochlear,*) their shell rather fragile, almost without lamellæ; upper valve concave, colour rather deep violet, form variable.

"Besides these three kinds of Molluscæ, of which the number was considerable, several species of calcareous polypi were found; but those which could be obtained were too imperfect to allow of their being correctly described.

"The iron which was used to protect the copper on the bottom of the Constance frigate having been subjected to chemical analysis, the following are the results.

"This iron, which was in small fragments, very friable, little attracted by the loadstone, soft to the touch, and soiling the fingers like plumbago, gave out in rubbing it a very strong smell, very much like that of burnt linseed oil. Its colour on the exterior was a brownish yellow, and its interior a blackish grey, studded with little points extremely brilliant.

"A short time after they had been taken from the keel of the frigate, where they were covered with a layer of hydrated peroxide of iron, of six or eight lines in thickness, and been enclosed in a paper box, these fragments became strongly heated, and underwent a real combustion by means of the oxygen of the atmosphere; the combustion was accompanied by the production of a certain quantity of aqueous vapour.

"In order to ascertain whether this elevation of temperature was really alone owing to the absorption of the oxygen, a case containing twelve *grammes* of this iron was placed under a receiver, which contained two hundred millimetres, inverted over a tube of mercury; and it was observed, in the course of an hour, that this air had diminished by forty millimetres, or one-fifth of its volume. Examining afterwards that which remained in the receiver, it was discovered that it had no effect whatever either on lime-water or the tincture of *tournesol*,—that it was not inflammable,—that it extinguished a candle; in a word, that it presented all the negative qualities of azotic gas, strongly infected with the smell before stated.

"It must be evident that the oxygen which was absorbed in this experiment was employed solely in burning the iron, which was already in a state of *protoxide*, as was indicated by its little degree of cohesion, by the avidity with which it seized this principle, and by its dissolving in sulphuric acid, which operated without effervescence, and without disengaging hydrogen gas.

"Five *grammes* of this oxidized iron being reduced to an impalpable powder, and then made red-hot in a platina crucible, and mixed with three parts of *potasse à l'alcool*, were reduced to a clammy mass, coloured on its edges with a clear beautiful green, and with a greenish yellow on the other parts; which at once indicated the presence of a small portion of manganese, and that of a little *chrôme;* metals which are found united in almost all sorts of iron. Treated in the usual way, this mass exhibited—

"First, Traces, scarcely sensible, of these two metals.

"Secondly, One *gramme* of brilliant black powder, soft to the touch, staining paper, insoluble in muriatic acid when applied boiling: it was therefore a true percarburet of iron.

"Thirdly, Three *grammes* and ten *decigrammes* of peroxide of iron.

"On being subjected to the action of boiling water, five grammes of this pulverized iron gave out three *decigrammes* of soluble matter, composed, for the greater part, of *hydrochlorate* of iron, and a trace of hydrochlorate of magnesia, together with a little organic matter, the combination of which with the iron will account for the insufferable smell which it gave out when the iron was heated. This saline solution sensibly reddened the litmus paper: an effect which was owing to the muriatic acid, which, in uniting with oxidized iron, and with most other metallic oxides, never forms combinations which are perfectly neutral, but which are always more or less acid.

"It has in vain been attempted to discover in this oxidized iron the presence of silex, of alumine, and of the sulphuric and carbonic acids, either free, or in combination.

"It results from this analysis, that the fragments of the protectors, which have been the object of it, are composed, in a hundred parts, of about

64 oxidized iron,

20 of plumbago, or percarburet of iron,

6 of matter soluble in water, hydrochlorate of magnesia, hydrochlorate of iron, hydrochlorate of soda, hydrochlorate of magnesia, and organic matter, and

10 of water; as in fragments pulverised and heated for half an hour at a temperature of 100°, they lost 1-10th of their weight.

"As to the reddish yellow matter, with small protuberances like nipples, which formed a thick layer on the surface of the protectors, it was formed of 75 parts of oxide of iron at the most, and 25 parts of water, besides some atoms of hydrochlorate of iron, hydrochlorate of soda, and hydrochlorate of magnesia."

Had not the health of Davy unfortunately declined at the very period when his energies were most required, such is the unbounded confidence which all must feel in his unrivalled powers of vanquishing practical difficulties, and of removing the obstacles which so constantly thwart the applications of theory, that little doubt can be entertained but he would soon have discovered some plan by which the adhesion of marine bodies to the copper sheathing might have been prevented, and his principle of Voltaic protection thus rendered available. An experiment indeed, altogether founded upon this same principle, has been already proposed, and will be shortly tried in the British navy, by building a schooner, and fastening its materials together with copper bolts, and afterwards sheathing the bottom with thin plates of iron, which are to be protected by bands of zinc. At the same time, another schooner is to be built, in which the fastenings are to consist entirely of iron bolts and nails, the former to be protected by a zinc ring under each head or clench, and the latter to have a small piece of zinc soldered under its head.

This plan of protection was first adopted in America, at the recommendation of Dr. Revere; and upon its successful issue, that gentleman was lately induced to take out letters patent not only in England, but in all the maritime countries of Europe, for the sole right of manufacturing iron sheathing, bolts, and nails, thus protected.

As no doubt now exists as to the principle of the protection of iron by zinc, the bolts and nails may be expected to remain free from rust as long as the more oxidable metal lasts; but with regard to the success of the iron sheathing, it is impossible to entertain the same confidence; for what, in this case, is to prevent the adhesion of shell-fish and sea-weed upon its surface? Let it be remembered, that it is only when the copper is in the act of solution in sea-water that the sheathing remains clean. In the year 1829, the Tender to the Flag-ship at Plymouth had her copper on one side of the bottom painted with white lead: in six months, this side was covered with long weeds; while the other side, which had been left bright, and consequently exposed to the solvent action of the salt-water, was found entirely free from all such adhesions.

CHAPTER XIV.

The failure of the Ship protectors a source of great vexation to Davy.—His Letters to Mr. Poole.—He becomes unwell.—He publishes his Discourses before the Royal Society.—Critical Remarks —and Quotations.—He goes abroad in search of health.—His Letter to Mr. Poole from Ravenna.—He resigns the Presidency of the Royal Society.—Mr. Gilbert elected *pro tempore.*—Davy returns to England, and visits his friend Mr. Poole.—Salmonia, or Days of Fly fishing.—An Analysis of the Work, with various extracts to illustrate its character.

THE friends of Sir Humphry Davy saw with extreme regret that the failure of his plan for protecting copper sheathing had produced in his mind a degree of disappointment and chagrin wholly inconsistent with the merits of the question; that while he became insensible to the voice of praise, every nerve was jarred by the slightest note of disapprobation. I apprehend, however, that the change of character which many ascribed to the mortification of wounded pride, ought in some measure to be referred to a declining state of bodily power, which had brought with it its usual infirmities of petulance and despondency. The letters I shall here introduce may perhaps be considered as indicating that instinctive desire for quiet and retirement which frequently marks a declining state of health, and they will be followed by others of a less equivocal character.

TO THOMAS POOLE, ESQ.

MY DEAR POOLE, Grosvenor Street, Nov. 24, 1824.

It is very long since I have heard from you, Mr. A——, whom you introduced to me, has sometimes given me news of you, and I have always heard of your health and well-being with pleasure.

My immediate motive for writing to you now is somewhat, though not entirely selfish. You know I have always admired your neighbourhood, and I have lately seen a place advertised there, called, I think, ——, not far from Quantock, and combining, as far as advertisement can be trusted, scenery, fishing, shooting, interest for money, &c.

If it is not sold, pray give me a little idea of it; I have long been looking out for a purchase,—perhaps this may suit me. After all, it may be sold; if so, no harm is done.

I go on labouring for utility, perhaps more than for glory; caring something for the judgment of my contemporaries, but more for that of posterity; and confiding with boldness in the solid judgment of Time.

I have lately seen some magnificent country in the Scandinavian peninsula, where Nature, if not a kind, is at least a beautiful mother.—I wonder there have not been more poets in the North.

I am, my dear Poole,

Very affectionately your old friend,

H. Davy.

TO THE SAME.

MY DEAR POOLE, January 5, 1825.

MY proposition to come into Somersetshire about the 10th was founded upon two visits which I had to pay in this county, Hants; I am now only about sixty miles from you; and had you been at home, I should have come on to Nether Stowey The 13th is the first meeting of the Royal Society after the holidays; and though I might do my duty by deputy, yet I feel that this would not be right, and I will not have the honour of the chair without conscientiously taking the labours which its possession entails. I regret therefore that I cannot be with you next week. God bless you.

Believe me always, my dear Poole,
Your affectionate friend,
H. DAVY.

TO THE SAME.

MY DEAR POOLE, Park Street, Feb. 11, 1825.

* * * * * *

I had a letter a few days ago from C——, who writes in good spirits, and who, being within a few miles of London, might, as far as his friends are concerned, be at John a Grot's house. He writes with all his ancient power. I had hoped that, as his mind became subdued, and his imagination less vivid, he might have been able to apply himself to persevere, and to give to the world some of those trains of thought, so original, so impressive, and at which we have so often wondered.

I am writing this letter at a meeting of the Trustees of the British Museum, which will account for its want of correction. Lest I should be more desultory, I will conclude by subscribing myself, my dear Poole, your old and affectionate friend,

H. DAVY.

TO THE SAME.

MY DEAR POOLE, Feb. 28, 1825.

I AM very much obliged to you for your two letters, which I received in proper time. I have deferred writing, in the hopes that I might be able to pay you a visit and see the property, but I now find this will be impossible. I have a cold, which has taken a stronger and more inflammable character than usual, which obliges me to lay myself up; and in this weather it would be worse than imprudent to travel.

I have seen Mr. Z——, and can perfectly re-echo your favourable sentiments respecting him. I saw the plan of the estate, and heard every thing he had to say respecting the value, real and imaginary, of the lands. He certainly hopes at this moment for a fancy price, and he is right if he can get it.

* * * * * *

I have less fancy for the place, from finding the trout-stream a brook in summer, where salmon-trout, or salmon, could not be propagated; for one of my favourite ideas in a country residence is varied and multiplied experiments on the increase and propagation of fish.

What I should really like would be a place with a couple of hundred acres of productive land, and plenty of moor, a river running through it, and the sea before it; and not farther from London than

Hampshire—a day's journey. There are such places along the coast, though perhaps in my lifetime they will not be disposed of. I should also like to be within a few miles of you; for it is one of the regrets in the life which I lead, that devotion to the cause of science separates me very much from friends that I shall ever venerate and esteem. God bless you, my dear Poole,

Very affectionately yours,

H. DAVY.

TO THE SAME.

MY DEAR POOLE, Pixton near Dulverton, Nov. 1, 1826.

I CANNOT be in your neighbourhood, without doing my best to see you; and it is my intention to come to Stowey on Sunday. I hope I shall find you at home, and quite well.

Mr. T——, who is here, gives me a very good account of you, which I trust I shall be personally able to verify.

If you are at leisure, I will try to shoot a few woodcocks on Monday on the Quantock hills; on Tuesday I must go east.

I have not been well lately. I cannot take the exercise which twenty years ago I went lightly and agreeably through. Will you have the kindness to hire a pony for me, that I may ride to your hills?

I am sorry I did not know of your journey to Ireland and Scotland. I was in both those countries at the time you visited them, and should have been delighted to have met you.

Do not write to me; for, even if you should not

be at home, Stowey is not more than ten or twelve miles out of my way; but I hope I shall find you.

I am, my dear Poole,

Your old and sincere friend,

H. Davy.

The complaints, as to the loss of his strength, which are expressed in the preceding letter, were but too well founded. Mr. Poole informs me that, during this visit in 1826, it was affecting to observe the efforts he made to continue his field sports. From being unable to walk without fatigue, he was compelled to have a pony to take him to the field, from which he dismounted only on the certainty of immediate sport.

On his return to London, his indisposition increased: he complained to me of palpitation of the heart, and of an affection in the trachea, which led him to fear that he might be suffering under the disease of which his father died.

The fatigue attendant upon the duties of the anniversary of the Royal Society (November 30th) completely exhausted him; and after his re-election as President, he was reluctantly obliged to retire, and to decline attending the usual dinner upon that occasion.

In January 1827, Sir Humphry Davy published the Discourses which he had delivered before the Royal Society, at six successive anniversary meetings, on the award of the Royal and Copley medals. They were published in compliance with a resolution of a meeting of the Council, held on the 21st of December 1826.

The practice of delivering an annual oration before the Royal Society, on the occasion of presenting the medal upon Sir Godfrey Copley's donation, prevailed during the presidency of Sir John Pringle; it was, however, during a long interval discontinued, and only revived during the latter years of Sir Joseph Banks.

The discourse usually commenced with a short tribute of respect to the memory of those distinguished Fellows who had died since the preceding anniversary. It then proceeded to announce the choice of the Council in its award of the medals, enumerating the objects and merits of the several communications which had been honoured with so distinguished a mark of approbation, and stating the circumstances which had directed the judges in their decision.

Much has been said and written upon the inutility, and even upon the mischievous tendency of this practice; and great stress has been laid upon the vices inseparably connected, as it is asserted, with the style of composition to which it gives origin. It appears to me, however, that it is only against the meretricious execution, not against the temperate use of such discourses, that this charge can be fairly and consistently sustained; and in the chaste and yet powerful addresses of Davy, such an opinion will find its best sanction, and obtain its strongest support.

Does it follow, because praise, when unduly lavished upon the labours of the scientific dead, may create comparisons and preferences injurious to the living, that we are to stifle the noblest aspirations of our nature, and become as cold and silent as the grave that encloses their remains? Does it follow, because an undisciplined ardour may have occasionally exaggerated the merits of our contemporaries, that we are henceforth to withhold from them a just tribute of applause at their discoveries—to forego the advantages which science must derive from a plan so well calculated to awaken the flagging attention, to infuse into stagnant research a renewed spirit of animation, and to encourage the industry of the labourer in the abstract regions of science, with prospects gleaming with sunshine, and luxuriant in the fruitfulness which is to reward him?

Such was the character, such the effect of Davy's discourses. They exhibit a great assemblage of diversified talents, and display the refined views he entertained with respect to the mutual relations which the different sciences maintain with each other; they evince, moreover, a great command of language, and a power to give exact expression to what his mind had conceived.

To these six Discourses is prefixed his Address upon taking the chair of the Royal Society for the first time; the subject of which is "The present state of that Body, and the Progress and Prospects of Science." Upon this occasion, he particularly adverts to the light which the different branches of science may reflect upon each other. "In pure Mathematics—though their nature, as a work of intellectual combination, framed by the highest efforts of human intelligence, renders them incapable of receiving aids from observations of external phenomena, or the invention of new instruments, yet they are, at this moment, abundant in the promise of new applications; and many of the departments of philosophical enquiry which appeared formerly to bear no relation to quantity, weight, figure, or number, as I shall more particularly mention hereafter, are now brought under the dominion of that sublime science, which is, as it were, the animating principle of all the other sciences."

"In the theory of light and vision, the researches of Huygens, Newton, and Wollaston, have been followed by those of Malus; and the phenomena of polarization are constantly tending to new discoveries; and it is extremely probable that those beautiful results will lead to a more profound knowledge than has hitherto been obtained, concerning the intimate constitution of bodies, and establish a near connexion between mechanical and chemical philosophy."

"The subject of heat, so nearly allied to that of light, has lately afforded a rich harvest of discovery; yet it is fertile in unexplored phenomena. The question of the materiality of heat will probably be solved at the same time as that of the undulating hypothesis of light, if, indeed, the human mind should ever be capable of understanding the causes of these mysterious phenomena. The applications of the doctrine of heat to the atomic or corpuscular philosophy of chemistry abound in new views, and probably at no very distant period these views will assume a precise mathematical form."

"In Electricity, the wonderful instrument of Volta has done more for the obscure parts of physics and chemistry, than the microscope ever effected for natural history, or even the telescope for astronomy. After presenting to us the most extraordinary and unexpected results in chemical analysis, it is now throwing a new light upon magnetism.

'Suppeditatque novo confestim lumine lumen.'

"I must congratulate the Society on the rapid progress made in the theory of definite proportions, since it was advanced in a distinct form by the ingenuity of Mr. Dalton. I congratulate the Society on the promise it affords of solving the recondite changes, owing to motions of the particles of matter, by laws depending upon their weight, number, and figure, and which will be probably found as simple in their origin, and as harmonious in their relations, as those which direct the motions of the heavenly bodies, and produce the beauty and order of the celestial systems.

"The crystallizations, or regular forms of inorganic matter, are intimately connected with definite proportions, and depend upon the nature of the combinations of the elementary particles; and both the laws of electrical polarity, and the polarization of light, seem related to these phenomena. As to the origin of the primary arrangement of the crystalline matter of the globe, various hypotheses have been applied, and the question is still agitated, and is perhaps above the present state of our knowledge; but there are two principal facts which present analogies on the subject,—one, that the form of the earth is that which would result, supposing it to have been originally fluid; and the other, that in lavas, masses decidedly of igneous origin, crystalline substances, similar to those belonging to the primary rocks, are found in abundance."

It is the privilege of genius to be in advance of the age, and to see, " as by refraction, the light, as yet below the horizon." It is with such a feeling that I have introduced the foregoing extracts, which I cannot but regard as prophetic of future discoveries.

The first discourse was delivered on the 30th of November 1821, on the occasion of announcing the award of two medals, on Sir Godfrey Copley's donation; one to J. F. W. Herschel, Esq. for his various papers on mathematical and physico-mathematical subjects; and the other, to Captain Edward Sabine, R.A., for his papers containing an account of his various experiments and observations made during a voyage and expedition in the Arctic regions.

As I am desirous that the reader should be made acquainted with the nature and style of the address with which he accompanied the presentation of the medal, I cannot select a happier example, or one in the sentiments of which every person will more readily participate, than the following:—

" MR. HERSCHEL—Receive this medal, Sir, as a mark of our respect, and of our admiration of those talents which you have applied with so much zeal and success, and preserve it as a pledge of future exertions in the cause of Science and of the Royal Society; and, believe me, you can communicate your labours to no public body by whom they will be better received, or through whose records they will be better known to the philosophical world. You are in the prime of life, in the beginning of your career, and you have powers and acquirements capable of illustrating and extending every branch of physical enquiry; and, in the field of science, how many are the spots not yet cultivated! Where the laws of sensible become connected with those of insensible motions, the mechanical with the chemical phenomena, how little is known! In electricity, magnetism, in the relations of crystallized forms to the weight of the elements of bodies, what a number of curious and important objects of research! and they are objects which you are peculiarly qualified to pursue and illustrate.

"May you continue to devote yourself to philosophical pursuits, and to exalt your reputation, already so high—

'Virtutem extendere factis.'

And these pursuits you will find not only glorious, but dignified, useful, and gratifying in every period of life: this, indeed, you must know best in the example of your illustrious father, who, full of years and of honours, must view your exertions with infinite pleasure; and who, in the hopes that his own imperishable name will be permanently connected with yours in the annals of philosophy, must look forward to a double immortality."

In the discourse of the succeeding year, it was his painful duty to announce the death of the elder Herschel, whom, in his former address, he had eulogized in such eloquent and touching language.

In alluding to the labours and discoveries of Sir William Herschel, he observed, that "they have so much contributed to the progress of modern astronomy, that his name will probably live as long as the inhabitants of this earth are permitted to view the solar system, or to understand the laws of its motions. The world of science—the civilized world, are alike indebted to him who enlarges the boundaries of human knowledge, who increases the scope of intellectual enjoyment, and exhibits the human mind in possession of new and unknown powers, by which it gains, as it were, new dominions in space; acquisitions which are imperishable—not like the boundaries of terrestrial states and kingdoms, or even the great monuments of art, which, however extensive and splendid, must decay—but secured by the grandest forms and objects of nature, and registered amongst her laws."

One more quotation, and I shall conclude with the conviction that the splendid specimens I have adduced must fully justify the opinion already offered as to the taste, power, and eloquence with which, as President of the Royal Society, he discharged the most delicate and arduous of all its duties.

In his address to Mr. (now Dr.) Buckland, on delivering to him the medal for his important memoir on the fossile remains discovered in the cave near Kirkdale, he thus concludes:—"If we look with wonder upon the great remains of human works, such as the columns of Palmyra, broken in the midst of the desert; the temples of Pæstum, beautiful in the decay of twenty centuries; or the mutilated fragments of Greek sculpture in the Acropolis of Athens, or in our own Museum, as proofs of the genius of artists, and power and riches of nations now past away; with how much deeper a feeling of admiration must we consider those grand monuments of nature which mark the revolutions of the globe—continents broken into islands; one land produced, another destroyed; the bottom of the ocean become a fertile soil; whole races of animals extinct, and the bones and exuviæ of one class covered with the remains of another; and upon these graves of past generations—the marble or rocky tombs, as it were, of a former animated world, new generations rising, and order and harmony established, and a system of life and beauty produced, as it were, out of chaos and death, proving the infinite power, wisdom, and goodness of the Great Cause of all being!"

I have noticed the apparent commencement of that general indisposition which had for some time been stealing upon him, undermining his powers, oppressing his spirits, and subduing his best energies; but in the end of 1826, his complaint assumed a more decided and alarming form. Feeling more than usually unwell, while on a visit to his friend Lord Gage, he determined to return to London, and was seized while on his journey, at Mayersfield, with an apoplectic attack. Prompt and copious bleeding, however, on the spot, arrested the symptoms more immediately threatening the extinction of life, and enabled him to reach home; but paralysis, the usual consequence of such seizures, had obviously, though at first but slightly, diminished his muscular powers, and given an awkwardness to his gait.

As soon as the more immediate danger of the attack had passed away, it was thought expedient to recommend, as the best means of his farther recovery, a residence in the southern part of Europe, where he would be removed from all the cares and anxieties that were inseparably connected with his continuance in London; and he accordingly quitted England, with the intention of spending what remained of the winter in Italy.

The following interesting letter to his friend will sufficiently explain the serious character of his malady, and the degree of bodily infirmity which accompanied it.

TO THOMAS POOLE, ESQ.

MY DEAR POOLE, Ravenna, March 14, 1827.

I SHOULD have answered your letter immediately, had it been possible; but I was, at the time I received it, very ill, in the crisis of the complaint under which I have long suffered, and which turned out to be a determination of the blood to the brain; at last producing the most alarming nervous symptoms, and threatening the loss of power and of life.

Had I been in England, I should gladly have promoted the election of your friend at the Athenæum: your certificate of character would always be enough for me; for, like our angling evangelical

Isaac Walton, I know you choose for your friends only good men.

I am, thank God, better, but still very weak, and wholly unfit for any kind of business and study. I have, however, considerably recovered the use of all the limbs that were affected ; and as my amendment has been slow and gradual, I hope in time it may be complete : but I am leading the life of an anchorite, obliged to abstain from flesh, wine, business, study, experiments, and all things that I love ; but this discipline is salutary, and for the sake of being able to do something more for science, and, I hope, for humanity, I submit to it ; believing that the Great Source of intellectual being so wills it for good.

I am here lodged in the Apostolical Palace, by the kindness of the Vice-Legate of Ravenna, a most admirable and enlightened prelate, and who has done every thing for me that he could have done for a brother.

I have chosen this spot of the declining empire of Rome, as one of solitude and repose, as out of the way of travellers, and in a good climate ; and its monuments and recollections are not without interest. Here Dante composed his divine works. Here Byron wrote some of his best and most moral (if such a name can be applied) poems ; and here the Roman power that began among the mountains with Romulus, and migrated to the sea, bounding Asia and Europe under Constantine, made its last stand, in the marshes formed by the Eridanus, under Theodorick, whose tomb is amongst the wonders of the place.

After a month's travel in the most severe weather I ever experienced, I arrived here on the 20th of February. The weather has since been fine. My brother and friend, who is likewise my physician, accompanied me; but he is so satisfied with my improvement, as to be able to leave me for Corfu; but he is within a week's call.

I have no society here, except that of the amiable Vice-Legate, who is the Governor of the Province; but this is enough for me, for as yet I can bear but little conversation. I ride in the pine forest, which is the most magnificent in Europe, and which I wish you could see. You know the trees by Claude Lorraine's landscapes: imagine a circle of twenty miles of these great fan-shaped pines, green sunny lawns, and little knolls of underwood, with large junipers of the Adriatic in front, and the Apennines still covered with snow behind. The pine wood partly covers the spot where the Roman fleet once rode,—such is the change of time!

It is my intention to stay here till the beginning of April, and then to go to the Alps, for I must avoid the extremes of heat and cold.

God bless you, my dear Poole. I am always your old and sincere friend, H. DAVY.

Feeling that his recovery was tardy, and that perfect mental repose was more than ever necessary for its advancement, he determined to resign the chair of the Royal Society; and he accordingly announced that intention, by a letter to his friend Mr. Davies Gilbert, Vice-President of the Society.

TO DAVIES GILBERT, ESQ. M. P. V. P. R. S. &c. &c.

MY DEAR SIR, Salzburgh, July 1, 1827.

YESTERDAY, on my arrival here, I found your two letters. I am sorry I did not receive the one you were so good as to address to me at Ravenna; nor can I account for its miscarriage. I commissioned a friend there to transmit to me my letters from that place after my departure, and I received several, even so late as the middle of May, at Laybach, which had been sent to Italy, and afterwards to Illyria. I did not write to you again, because I always entertained hopes of being able to give a better account of the state of my health. I am sorry to say the expectations of my physicians of a complete and rapid recovery have not been realized. I have gained ground, under the most favourable circumstances, very slowly; and though I have had no new attack, and have regained, to a certain extent, the use of my limbs, yet the tendency of the system to accumulate blood in the head still continues, and I am obliged to counteract it by a most rigid vegetable diet, and by frequent bleedings with leeches and blisterings, which of course keep me very low. From my youth up to last year, I had suffered, more or less, from a slight hemorrhoidal affection; and the fulness of the vessels, then only a slight inconvenience, becomes a serious and dangerous evil in the head, to which it seems to have been transferred. I am far from despairing of an ultimate recovery, but it must be a work of time, and the vessels which have been over distended only very slowly regain their former dimensions and tone:

and for my recovery, not only diet and regimen and physical discipline, but a freedom from anxiety, and from all business and all intellectual exertion, is absolutely required.

Under these circumstances, I feel it would be highly imprudent and perhaps fatal for me, to return, and to attempt to perform the official duties of President of the Royal Society. And as I had no other feeling for that high and honourable situation, except the hope of being useful to society, so I would not keep it a moment without the security of being able to devote myself to the labour and attention it demands. I beg therefore you will be so good as to communicate my resignation to the Council and to the Society at their first meeting in November, after the long vacation; stating the circumstances of my severe and long continued illness, as the cause. At the same time, I beg you will express to them how truly grateful I feel for the high honour they have done me in placing me in the chair for so many successive years. Assure them that I shall always take the same interest in the progress of the grand objects of the Society, and throughout the whole of my life endeavour to contribute to their advancement, and to the prosperity of the body.

Should circumstances prevent me from sending, or you from receiving any other communication from me before the autumn (for nothing is more uncertain than the post in Austria, as they take time to read the letters), I hope this, which I shall go to Bavaria to send, will reach you safe, and will be sufficient to settle the affair of resignation.

It was my intention to have said nothing on the subject of my successor. I will support by all the means in my power the person that the leading members of the Society shall place in the chair; but I cannot resist an expression of satisfaction in the hope you held out, that an illustrious friend of the Society, illustrious from his talents, his former situation, and, I may say, his late conduct, is likely to be my successor.

I wish my name to be in the next Council, as I shall certainly return, *Deo volente,* before the end of the session, and I may, I think, be of use; and likewise, because I hope it may be clearly understood that my feelings for the Society are, as they always were, those of warm attachment and respect. Writing still makes my head ache, and raises my pulse. I will therefore conclude, my dear Sir, in returning you my sincere thanks for the trouble you have had on my account, and assuring you that I am

Your obliged and grateful friend and servant,

H. DAVY.

Pray acknowledge the receipt of this letter, by addressing me, "*poste restante,* Laybach, Illyria, Austria;" and let me know if Mr. Hudson is still Assistant-Secretary, and where Mr. South is. I send this letter from Frauenstein, Bavaria, July 2, that it may not be opened, as all my letters were at Salzburgh. There was one of them must have *amused* Prince Metternich, on the state of parties in England, from a Member of the Upper House.

In consequence of this letter, the Council of the Royal Society, by a resolution passed at a very full meeting held on the 6th of November, 1827, appointed Mr. Davies Gilbert to fill the chair, until the general body should elect a President, at the ensuing anniversary.

The following letter will show his subsequent course of proceeding.

TO THOMAS POOLE, ESQ.

Park Street, Grosvenor Square,
Oct. 29, 1827.

MY DEAR POOLE,

I HOPE you received a letter which I addressed to you from Ravenna in the spring. It was my intention to have returned to Italy from the Alpine countries, where I spent the summer; but my recovery has been so slow, and so much uneasiness in the head and weakness in the limbs remained in September, that I thought it wiser to return to my medical advisers in London.

I have consulted all the celebrated men who have written upon or studied the nervous system. They all have a good opinion of my case, and they all order absolute repose for at least twelve months longer, and will not allow me to resume my scientific duties or labours at present; and they insist upon my leaving London for the next three or four months, and advise a residence in the west of England. Now, my dear friend, you recollect our conversation upon the subject of a residence—I think Mr. C.'s is not very far from you. Pray let me know something on this head. I want very little of any thing, for I am almost on a vegetable diet; and a little horse exercise, a very little shooting, and a little quiet society, are what I am in search of, with some facilities of procuring books. I have thought of Minehead, Ilfracombe, Lymouth, and Penzance; but I have not yet determined the point.

Horse exercise and shooting are necessary to bring back my limbs to their former state, and therefore Bath and Brighton will not do for me. God bless you, my dear Poole, and pray let me hear from you. Your affectionate, H. DAVY.

P.S. I hope you got the copy of my discourses.

TO THE SAME.

MY DEAR POOLE, Firle, near Lewes, Nov. 4, 1827.

I HAVE this moment received your very kind and most friendly letter. I have made my first visit to my friend Lord G——, where I was taken ill last year; and have borne the journey well, and have enjoyed the small society here; but I am very weak indeed, and I cannot yet walk more than a mile. One of three plans, I shall hope to adopt; two of them you have most amiably suggested, the other is to go to Penzance. My only objection to the last is the fear of too much society. Whatever I do, I will first come to you and take your advice.

When I returned, I had little hopes of recovery; but the assurances of my physicians that I may again, with care, be re-established, have revived me, and I have certainly gained ground, and gained strength, by the plan I am now pursuing.

As soon as I return to London, I will write to you. If I can find a companion, I think Mr. C——'s house will do admirably; but I must see it, as a temperate situation is a *sine quâ non.*

I need not say how grateful I am for your kindness, and if I recover, how delighted I shall be to owe the means to so excellent and invaluable a friend. God bless you.

I am, my dear Poole, your affectionate,

H. DAVY.

TO THE SAME.

MY DEAR POOLE, Firle, Nov. 7.

I AM going to London to-morrow, and after staying two or three days, to try a new plan of medical treatment, which my physicians recommend, I shall come westward, and profit by your kindness, and adopt whichever of the plans will promise to be most salutary.

If I take Mr. C——'s house, Lady Davy will come to me. With respect to society, I want only a friend, or one person or two at most, to prevent entire solitude, and I am too weak to bear much conversation, and wholly unfit to receive any but persons with whom I am in the habits of intimacy.

I can hardly express to you how deeply I feel your kindness.

* * * * * *

As I must travel slowly, I shall not probably be at Stowey before Wednesday or Thursday next. Pray do not ask any body to meet me. I am upon the *strictest* diet,—a wing of a chicken and a plain rice or bread pudding is the extreme of my *gourmanderie*. God bless you.

My dear Poole, your affectionate friend,

H. DAVY.

Mr. Poole has been so obliging as to communicate to me some interesting particulars connected with the visit to which the foregoing letters allude.

"During this last vist, (November and December 1827,) his bodily infirmity was very great, and his sensibility was painfully alive on every occasion. Unhappily, he had to sustain the affliction of the sudden death of Mr. R——, the son of a friend whom he highly valued; and though this afflicting event was, by the considerate and anxious attention of Lady Davy, first communicated to me by letter, to be imparted to him with every precaution, to avoid his being suddenly shocked, yet it was many days before he could recover his usual spirits, feeble as they were, and resume his wonted occupation.

"On his arrival, he said, 'Here I am, the ruin of what I was;' but nevertheless, the same activity and ardour of mind continued, though directed to different objects. He employed himself two or three hours in the morning on his *Salmonia*, which he was then writing; he would afterwards take a short walk, which he accomplished with difficulty, or ride. After dinner, I used to read to him some amusing book. We were particularly interested by Southey's Life of Nelson. 'It would give Southey,' he said, 'great pleasure, if he knew how much his narrative affected us.'*

"In the evening, Mr. and Mrs. W——, the former of whom he had long known, frequently came to make a rubber at whist. He was averse to see-

* His admiration of this work bursts forth in his *Salmonia*, which he was writing at that time. He styles it "an immortal monument raised by Genius to Valour."

ing strangers; but on being shown the drawings of Natural History of a friend of mine of great talent, Mr. Baker of Bridgwater, he was anxious to know him, and was much pleased with his company. He suggested to him various subjects for investigation, concerning insects, and fish, particularly the eel. What pleasure would it give him were he now alive, to learn the interesting result of those suggestions! I hope the public will soon be made acquainted with them.

"Natural History in general had been a favourite subject with him throughout his protracted illness; and during this last visit to me, he paid attention to that only; 'for,' said he, 'I am prohibited applying, and indeed I am incapable of applying, to any thing which requires severe attention."

"During the same visit, I remember his inherent love of the laboratory, if I may so express myself, being manifested in a manner which much interested me at the moment. On his visiting with me a gentleman in this neighbourhood, who had offered him his house, and who has an extensive philosophical apparatus, particularly complete in electricity and chemistry, he was fatigued with the journey, and as we were walking round the house very languidly, a door opened, and we were in the laboratory. He threw a glance around the room, his eyes brightened in the action, a glow came over his countenance, and he looked like himself, as he was accustomed to appear twenty years ago.

"You are aware that he was latterly a good shot, always an expert angler, and a great admirer of old Isaac Walton; and that he highly prided himself upon these accomplishments. I used to laugh at him, which he did not like; but it amused me to see such a man give so much importance to those qualifications. He would say, 'It is not the sport only, though there is great pleasure in successful dexterity,* but it is the ardour of pursuit, pure air, the contemplation of a fine country, and the exercise—all tend to invigorate the body, and to excite the mind to its best efforts."

" These amusements seemed to become more and more important in his estimation, as his health declined. It was affecting to observe the efforts he

* Mr. Children has just communicated to me the following amusing anecdote, which may be adduced in illustration of the delight he took in that sporting dexterity to which he alludes in the above passage. Davy, with a party of friends, had been engaged for several hours in fishing for pike, but very unsuccessfully; our philosopher gave up the sport in despair, but his companions having determined to try some more propitious spots, left him to his contemplations. About an hour afterwards, Mr. Children, on returning to his friend, saw him at a distance seated upon a gate, and apparently lashing the air with his fishing-line. What could be his object? As soon as Mr. Children came sufficiently near to make a signal, Davy, by his gestures, earnestly entreated him to keep away, while he continued his mysterious motions. At length, however, Mr. Children's patience was exhausted, and he walked up to him. " Was ever any thing more provoking!" exclaimed Davy; " if you had only remained quiet another minute I should have caught him—it is most vexatious!" " Caught what?" asked Mr. Children. " A dragon-fly," *(Libellula,)* answered Davy. " During your absence I have been greatly amused by watching the feeding habits of that insect, and having observed the eagerness with which they snapped up the little '*midges*,' I determined to arm my hook with one, and I can assure you I have had no small degree of sport; and had it not been for your unwelcome intrusion, I should most undoubtedly have captured one of them."

made to continue them with diminished strength. From being unable to walk without fatigue for many hours, he was, when he came to me in November 1826, obliged to have a pony to carry him to the field, from which he dismounted only on the certainty of immediate sport. In the following year, he could only take short and occasional rides to the covers, with his dogs around him, and his servant walking by his side and carrying his gun, but which I believe he never fired.

" During this visit, he more than once observed, ' I do not wish to live, as far as I am personally concerned; but I have views which I could develope, if it pleased God to save my life, which would be useful to science and to mankind.' "

Davy returned to town in December, and after an interval wrote the following letter:

TO THOMAS POOLE, ESQ.

Park Street, Grosvenor Square, Dec. 27, 1827.

MY VERY DEAR FRIEND,

I KNOW no reason why I have not written to you. It has been my intention every day, and I have been every day prevented by the sense of want of power, which is so painful a symptom of my malady.

I continue much as I was. My physicians augur well, and I have some repose in the hopes connected with the indefinite future. In the last twelvemonth, which I hope is a large portion, on the whole, of my purgatory expiation for crimes of commission or omission, the most cheerful, or rather the least miserable, days that I spent, were a good deal owing to your kindness, which I shall never forget. I would, if it were possible, make my letter something more than a mere bulletin of health, or the expression of the feelings of a sick man; but I can communicate no news. The papers will tell you more than is true; and our politicians seem ignorant of what they are to do at home, much more abroad.

* * * * * *

I have got for you a copy of my lectures on the Chemistry of Agriculture, which I shall send to you by the first opportunity. God bless you, my dear Poole. I am always your sincere,

Grateful, and affectionate friend,

H. DAVY.

In the letter which follows, Davy dwells upon a subject in Natural History, which appears to have greatly occupied his thoughts, and to have continued a predominant subject of his contemplation, even to the latest day of his life.

TO THOMAS POOLE, ESQ.

MY DEAR FRIEND, Park Street, January 1, 1828.

I WRITE to you immediately, because that part of your letter which relates to Mr. Baker's pursuits interests me very much: but before I begin on this subject, I will give you a short bulletin of the state of my health. I go on much as I did at Stowey, and my physicians have made no alterations in the plan of treatment: I am not worse, and they tell me I shall be better. Now for Mr. Baker—I am very glad that he is occupied with those enquiries. I am particularly anxious for information on the generation of eels; it is an unsolved problem since the time of Aristotle. I am sure that all eels come from the sea, where they are bred; but there may be one or two species or varieties of them. What Mr. Baker says about the difference between the common eel and the conger is well worthy of attention; but I have known changes more extraordinary than the obliteration, or destruction, of a small tubular member occasioned by difference of habits.

Were the salt-water eels and the fresh-water eels which he examined of the same size? Many individuals of various sizes should be examined to establish the fact of their specific difference. This would be the season for examining the genital organs of eels, for they breed in winter; and were I a little better, I should go to the sea for the purpose of making enquiries on the subject.

Sir Everard Home is firmly convinced that the animal is hermaphrodite and impregnates itself: this, though possible, appears to me very strange in so large an animal. If Mr. Baker will determine this point, I can promise him an immortality amongst our philosophical anglers and natural historians; and if he will give us the history of water-flies, imitated by fly-fishers, he will command our immediate gratitude.

Pray communicate this letter to him with my best wishes, and with my hopes that his talents, which are very great, will be applied to enlighten us.

I can give you no news; the weather is dreadful, and the blacks and yellows are descending in fog. I long for the fresh air of your mountains.

God bless you, my dear Poole. Many—many

happy new years to you. Pray remember me, with the compliments of this day, to your excellent neighbours at Stowey. Your affectionate friend,

H. DAVY.

TO THE SAME.

MY DEAR POOLE, Park Street, March 26.

YOUR letter has given me great pleasure; first, because you, who are an enlightened judge in such matters, approve of my humble contribution to agriculture; and, secondly, because it makes me acquainted with your kind feelings, health, and Mr. Baker's interesting pursuits.

Mr. Baker appears to me to have distinctly established the point that the eel and conger are of different species; and from his zeal and activity, I hope the curious problem of the generation of these animals will be solved. I shall expect with impatience the results of his enquiries.

Now for my health, my very dear friend. I wish I could speak more favourably; I certainly do not lose ground, but I am doubtful if I gain any; but I do not despair.

I am going, by the advice of my physicians, to try another Continental journey. If I get considerably better, I shall winter in Italy, where, in this case, I shall hope to see you, and where I shall have an apartment ready for you in Rome.

I have not been idle since I left your comfortable and hospitable house. I have finished my *Salmonia*, and sent it to the press.—"*Flumina amo sylvasque inglorius*."—I do not think you will be displeased with this little *jeu* of my sick hours.

Mr. A—— was very amiable in calling on me. There is nothing that annoys me so much in my illness as my helplessness in not being able to indulge in society.

Your grateful and affectionate friend,

H. Davy.

We will now, for a while, leave our philosopher to pursue his journey to Italy, while we take a review of his Salmonia; the first edition of which was published in the Spring of 1828. The second, and much improved edition,* from which I shall take my extracts, is dated from Laybach, Illyria, September 28, 1828, but which did not appear until 1829.

We are told in the preface, that these pages formed the occupation of the author during some months of severe and dangerous illness, when he was wholly incapable of attending to more useful studies, or of following more serious pursuits;—that they constituted his amusement in many hours, which otherwise would have been unoccupied and tedious;—and that they are published in the hope that they may possess an interest for those persons who derive pleasure from the simplest and most attainable kind of rural sports, and who practise the art, or patronise the objects of contemplation, of the philosophical angler.

He informs us that the conversational manner

* "Salmonia, or Days of Fly-fishing; in a series of Conversations; with some account of the habits of Fishes belonging to the genus Salmo. By an Angler. Second edition.—London, John Murray, 1829."

and discursive style were chosen as best suited to the state of health of the author, who was incapable of considerable efforts and long continued attention; and he adds, that he could not but have in mind a model, which has fully proved the utility and popularity of this method of treating this subject—"The Complete Angler," by Walton and Cotton.

The characters chosen to support these conversations, were HALIEUS, who is supposed to be an accomplished fly-fisher—ORNITHER, who is to be regarded as a gentleman generally fond of the sports of the field, though not a finished master of the art of angling—POIETES, who is to be considered as an enthusiastic lover of nature, and partially acquainted with the mysteries of fly-fishing; and PHYSICUS, who is described uninitiated as an angler, but as a person fond of enquiries in natural history and philosophy.

Such are the personages by whose aid the machinery is to be worked; but he tells us that they are of course imaginary, though the sentiments attributed to them, the author may sometimes have gained from recollections of real conversations with friends, from whose society much of the happiness of his early life had been derived; and he admits that, in the portrait of the character of HALIEUS, given in the last dialogue, a likeness will not fail to be recognised to that of the character of a most estimable physician, ardently beloved by his friends, and esteemed and venerated by the public.

The work is dedicated to Dr. Babington, "in remembrance of some delightful days passed in his society, and in gratitude for an uninterrupted friendship of a quarter of a century."

I am informed by Lady Davy, that the engravings of the fish, by which the work is illustrated, are from drawings of his own execution; so that he could not, like old Isaac Walton, "take the liberty to commend the excellent pictures to him that likes not the book, because they concern not himself."

It has frequently happened that, while works of deep importance have justly conferred celebrity upon the author, his minor productions have been entirely indebted to his name for their popularity, and to his authority for their value. This, however, cannot be said of Salmonia, for it possesses the stamp of original genius, and bears internal evidence of a talent flowing down from a very high source of intelligence. In a scientific point of view, it exhibits that penetrating observation by which a gifted mind is enabled to extract out of the most ordinary facts and every-day incidents, novel views and hidden truths; while it shows that a humble art (I beg pardon of the brothers of the Angle) may, through the skill of the master, be made the means of calling forth the affections of the heart, and of reflecting all the colours of the fancy. By regarding the work in relation to the history and condition of its author, it certainly acquires much additional interest. The familiar and inviting style of the dialogue, whenever he discusses questions of natural history, must convince us that he was as well calculated to instruct in the Lyceum, as we long since knew him to be to teach in the Academy.

Composed in the hour of sickness and prostration, the work displays throughout its composition a tone of dignified morality and an expansion of feeling, which may be regarded as in unison with a mind chastened but not subdued, and looking forward to a better state of existence. "I envy," says he, "no quality of the mind or intellect in others; be it genius, power, wit, or fancy: but if I could choose what would be most delightful, and I believe most useful to me, I should prefer a firm religious belief to every other blessing; for it makes life a discipline of goodness; creates new hopes, when all earthly hopes vanish; and throws over the decay, the destruction of existence, the most gorgeous of all lights; awakens life even in death, and from corruption and decay calls up beauty and divinity; makes an instrument of torture and of shame the ladder of ascent to paradise; and, far above all combinations of earthly hopes, calls up the most delightful visions of palms and amaranths, the gardens of the blest, the security of everlasting joys, where the sensualist and the sceptic view only gloom, decay, annihilation, and despair!"

While describing an animated scene of insect enjoyment, he bursts into an apostrophe, highly characteristic of that quick and happy talent for seizing analogies, which so eminently distinguished all his writings. I shall quote the passage.

"*Physicus.*—Since the sun has disappeared, the cool of the evening has, I suppose, driven the little winged plunderers to their homes; but see, there are two or three humble bees which seem languid with the cold, and yet they have their tongues still in the fountain of honey. I believe one of them is actually dead, yet his mouth is still attached to the flower. He has fallen asleep, and probably died whilst making his last meal of ambrosia.

" *Ornither.*—What an enviable destiny, quitting life in the moment of enjoyment, following an instinct, the gratification of which has been always pleasurable! so beneficent are the laws of Divine Wisdom.

" *Physicus.*—Like Ornither, I consider the destiny of this insect as desirable, and I cannot help regarding the end of human life as most happy, when terminated under the impulse of some strong energetic feeling, similar in its nature to an instinct. I should not wish to die like Attila, in a moment of gross sensual enjoyment; but the death of Epaminondas or Nelson, in the arms of victory, their whole attention absorbed in the love of glory, and of their country, I think really enviable.

" *Poietes.*—I consider the death of the martyr or the saint as far more enviable; for, in this case, what may be considered as a Divine instinct of our nature is called into exertion, and pain is subdued, or destroyed, by a secure faith in the power and mercy of the Divinity. In such cases, man rises above mortality, and shows his true intellectual superiority. By intellectual superiority, I mean that of his spiritual nature, for I do not consider the results of reason as capable of being compared with those of faith. Reason is often a dead weight in life, destroying feeling, and substituting, for principle, calculation and caution; and, in the hour of death, it often produces fear or despondency, and is rather a bitter draught than nectar or ambrosia in the last meal of life.

"*Halieus.*—I agree with Poietes. The higher and more intense the feeling under which death takes place, the happier it may be esteemed; and I think even Physicus will be of our opinion, when I recollect the conclusion of a conversation in Scotland. The immortal being never can quit life with so much pleasure as with the feeling of immortality secure, and the vision of celestial glory filling the mind, affected by no other passion than the pure and intense love of God."

We are not to suppose that, however soothing and consolatory such feelings and hopes may have been, they weaned him from the world, or diminished his natural love of life; on the contrary, no one would have more gratefully received the services of a Medea, as the following passage will sufficiently testify. "Ah! could I recover any thing like that freshness of mind which I possessed at twenty-five, and which, like the dew of the dawning morning, covered all objects and nourished all things that grew, and in which they were more beautiful even than in midday sunshine,—what would I not give! All that I have gained in an active and not unprofitable life. How well I remember that delightful season, when, full of power, I sought for power in others; and power was sympathy, and sympathy power;—when the dead and the unknown, the great of other ages and of distant places were made, by the force of the imagination, my companions and friends;—when every voice seemed one of praise and love;—when every flower had the bloom and odour of the rose; and every spray or plant seemed either the poet's laurel, or the civic oak—which appeared to offer themselves as wreaths to adorn my throbbing brow. But, alas! this cannot be.——"

After the example of the great Patriarch of Anglers, the author of Salmonia commences, through the assistance of the principal interlocutor of the dialogue, HALIEUS, to enumerate the delights of his art, and to vindicate it from the charge of cruelty.

"*Halieus.*—The search after food is an instinct belonging to our nature; and from the savage in his rudest and most primitive state, who destroys a piece of game, or a fish, with a club or spear, to man in the most cultivated state of society, who employs artifice, machinery, and the resources of various other animals, to secure his object, the origin of the pleasure is similar, and its object the same: but that kind of it requiring most art may be said to characterize man in his highest or intellectual state; and the fisher for salmon and trout with the fly employs not only machinery to assist his physical powers, but applies sagacity to conquer difficulties; and the pleasure derived from ingenious resources and devices, as well as from active pursuit, belongs to this amusement. Then, as to its philosophical tendency, it is a pursuit of moral discipline, requiring patience, forbearance, and command of temper. As connected with natural science, it may be vaunted as demanding a knowledge of the habits of a considerable tribe of created beings,—fishes, and the animals that they prey upon, and an acquaintance with the signs and tokens of the weather, and its changes, the nature of waters, and of the atmosphere. As to its poetical relations, it carries us into the most wild and beautiful scenery of nature; amongst the mountain lakes, and the clear and lovely streams that gush from the higher ranges of elevated hills, or that make their way through the cavities of calcareous strata. How delightful in the early spring, after the dull and tedious time of winter, when the frosts disappear and the sunshine warms the earth and waters, to wander forth by some clear stream, to see the leaf bursting from the purple bud, to scent the odours of the bank perfumed by the violet, and enamelled, as it were, with the primrose and the daisy; to wander upon the fresh turf below the shade of trees, whose bright blossoms are filled with the music of the bee; and on the surface of the waters to view the gaudy flies sparkling like animated gems in the sunbeams, whilst the bright and beautiful trout is watching them from below; to hear the twittering of the water birds, who, alarmed at your approach, rapidly hide themselves beneath the flowers and leaves of the water-lily; and, as the season advances, to find all these objects changed for others of the same kind, but better and brighter, till the swallow and the trout contend, as it were, for the gaudy May-fly, and till, in pursuing your amusement in the calm and balmy evening, you are serenaded by the songs of the cheerful thrush and melodious nightingale, performing the offices of paternal love, in thickets ornamented with the rose and woodbine."

On vindicating the pursuit from the charge of cruelty, he has advanced an argument that has not been commonly adduced upon this occasion. We have indeed all heard, that the operation of skinning is a matter of indifference to eels when they are used to it; but we are now told fish are so little annoyed by the hook, that though a trout has been hooked and played with for some minutes, he will often, after his escape with the artificial fly in his mouth, take the natural fly, and feed as if nothing had happened; having apparently learnt only from the experiment, that the artificial fly is not proper food. "I have caught pikes with four or five hooks in their mouths, and tackle which they had broken only a few minutes before; and the hooks seemed to have had no other effect than that of serving as a sort of *sauce piquante*, urging them to seize another morsel of the same kind."

Our author, however, takes a more special defence, by observing that, unlike old Isaac, he employs an artificial fly, instead of a living bait. Our notions about the cruelty of field sports is extremely capricious. Until the time of the Reformation, the canon law prohibited the use of the sanguinary recreations of hunting, hawking, and fowling, while the clergy, on account of their leisure, were allowed to exercise the harmless and humane art of angling. In later days, the indignation against this art has been excited by the supposed sufferings of the worm or bait, rather than by those of the fish; and thus far the author of Salmonia assumed a strong posture of defence; but he did not avail himself of all the advantages it commanded. He might have pleaded, that every fish he caught by his artificial fly was destined to prey upon an insect, and that by substituting a piece of silk for the latter, he would for every fish he might destroy, save from destruction many of those fairy beings that animate the air and sparkle in the sunbeam;—but it is, after all, folly to argue upon the subject of cruelty in our field sports. That animals should live by preying upon each other is the very basis of the scheme of creation; and in these days it is not necessary to expose the absurdities of the system of Samos and Indostan. Dr. Franklin, at one period of his life, entertained a sentimental abhorrence at eating any thing that had possessed life; and the reader may, perhaps, not object to be reminded of the manner in which he was cured of his prejudice. "I considered," says he, "the capture of every fish as a sort of murder, committed without provocation, since these animals had neither done, nor were capable of doing, the smallest injury to any one that should justify the measure. This mode of reasoning I conceived to be unanswerable. Meanwhile, I had formerly been extremely fond of fish; and when one of the cod was taken out of the frying-pan, I thought its flavour delicious. I hesitated some time between principle and inclination, till at last recollecting that, when the cod had been opened, some small fish were found in its belly, I said to myself, 'If you eat each other, I see no reason why we may not eat you,'—(His "wish was father to that thought")—I accordingly dined on the cod with no small degree of pleasure, and have since continued to eat like the rest of mankind."

Halieus is made to admit the danger of analysing too closely the moral character of any of our field sports; and yet, in the concluding chapter, he very unfairly and inconsistently attempts to ridicule the pursuit of a fox-hunter, "risking his neck to see the hounds destroy an animal which he preserves to be destroyed, and which is good for nothing." He who pursues a pleasure because it is rational, reasons because he cannot feel. "When the heart," says Sterne, "flies out before the understanding, it saves the judgment a world of pains."

Having, as the author thinks satisfactorily, settled the preliminary questions, HALIEUS, succeeds in persuading PHYSICUS to join him in fishing excursions; just as *Piscator* is represented by old Isaac, as having enlisted *Venator* into the brotherhood of the angle.

The dialogue now proceeds with great animation, during which the art and mystery of Piscatory tactics are unfolded with great skill; for the details of which the reader must be referred to the work itself. If, however, he be not already an angler, it may save him a world of pains to be informed, that to learn to fish by the book is little less absurd than "to make hay by the fair days in the almanack."

The manner in which he treats the various subjects of natural history necessarily connected with the pursuit is both amusing and instructive; and the whole work is studded and gemmed, as it were, with the most poetical descriptions.

In speaking of the swallow, *Poietes* exclaims—"I delight in this living landscape! The swallow is one of my favourite birds, and a rival of the nightingale; for he cheers my sense of seeing as much as the other does my sense of hearing. He is the glad prophet of the year, the harbinger of the best season:

he lives a life of enjoyment amongst the loveliest forms of nature: winter is unknown to him; and he leaves the green meadows of England in autumn, for the myrtle and orange groves of Italy, and for the palms of Africa: he has always objects of pursuit, and his success is secure. Even the beings selected for his prey are poetical, beautiful, and transient. The ephemeræ are saved by his means from a slow and lingering death in the evening, and killed in a moment, when they have known nothing of life but pleasure. He is the constant destroyer of insects—the friend of man; and, with the stork and the ibis, may be regarded as a sacred bird. His instinct, which gives him his appointed seasons, and teaches him always when and where to move, may be regarded as flowing from a divine source; and he belongs to the oracles of Nature, which speak the awful and intelligible language of a present Deity."

Poietes considers a full and clear river as the most poetical object in Nature.—"I will not fail to obey your summons. Pliny has, as well as I recollect, compared a river to human life. I have never read the passage in his works; but I have been a hundred times struck with the analogy, particularly amidst mountain scenery. The river, small and clear in its origin, gushes forth from rocks, falls into deep glens, and wantons and meanders through a wild and picturesque country, nourishing only the uncultivated tree or flower by its dew or spray. In this, its state of infancy and youth, it may be compared to the human mind, in which fancy and strength of imagination are predominant—it is more beautiful than useful.

When the different rills or torrents join, and descend into the plain, it becomes slow and stately in its motions; it is applied to move machinery, to irrigate meadows, and to bear upon its bosom the stately barge; — in this mature state it is deep, strong, and useful. As it flows on towards the sea, it loses its force and its motion, and at last, as it were, becomes lost, and mingled with the mighty abyss of waters."

Halieus adds—" One might pursue the metaphor still farther, and say, that in its origin—its thundering and foam, when it carries down clay from the bank, and becomes impure, it resembles the youthful mind, affected by dangerous passions. And the influence of a lake, in calming and clearing the turbid water, may be compared to the effect of reason in more mature life, when the tranquil, deep, cool, and unimpassioned mind is freed from its fever, its troubles, bubbles, noise, and foam. And, above all, the sources of a river—which may be considered as belonging to the atmosphere — and its termination in the ocean, may be regarded as imaging the divine origin of the human mind, and its being ultimately returned to, and lost in, the infinite and eternal Intelligence from which it originally sprang."

Halieus offers some curious observations with respect to the recollection of fish being associated with surrounding objects.

" I have known a fish that I have pricked retain his station in the river, and refuse the artificial fly, day after day, for weeks together; but his memory may have been kept awake by this practice, and the recollection seems local, and associated with surrounding objects; and if a pricked trout is chased into another pool, he will, I believe, soon again take the artificial fly. Or, if the objects around him are changed, as in autumn, by the decay of weeds, or by their being cut, the same thing happens; and a flood or a rough wind, I believe, assists the fly-fisher, not merely by obscuring the vision of the fish, but, in a river much fished, by changing the appearance of their haunts: large trouts almost always occupy particular stations, under, or close to, a large stone or tree; and probably, most of their recollected sensations are connected with this dwelling.

"*Physicus.*—I think I undertand you, that the memory of the danger and pain does not last long, unless there is a permanent sensation with which it can remain associated,—such as the station of the trout; and that the recollection of the mere form of the artificial fly, without this association, is evanescent.

"*Ornither.*—You are diving into metaphysics; yet, I think, in fowling, I have observed that the memory of birds is local. A woodcock that has been much shot at and scared in a particular wood, runs to the side where he has usually escaped, the moment he hears the dogs; but if driven into a new wood, he seems to lose his acquired habits of caution, and becomes stupid."

In alluding to the migrations of fishes, *Physicus* observes, "That he has always considered that the two great sources of change of place of animals, was the providing of food for themselves, and resting-places and food for their young. The great supposed migrations of herrings from the poles to the temperate zone appear to be only the approach of successive shoals from deep to shallow water, for the purpose of spawning. The migrations of salmon and trout are evidently for the purpose of depositing their ova, or of finding food after they have spawned."

In explaining the circumstances which render a migration into shallow water necessary for the developement of the ova, Davy evidently bears in mind the result of his very first experiment.*

"Carp, perch, and pike, deposit their ova in still water, in spring and summer, *when it is supplied with air by the growth of vegetables; and it is to the leaves of plants, which afford a continual supply of oxygen to the water, that the impregnated eggs usually adhere.*" Again: "Fish in spawning-time always approach great shallows, or shores covered with weeds, *which, in the process of their growth, under the influence of the sunshine, constantly supply pure air to the water in contact with them.*"

The following passage will afford a good specimen of the familiar dialogue, while it will convey to the reader some curious facts connected with the influence of sunshine.

"*Halieus.*—Well, gentlemen, what sport?

"*Poietes.*—The fish are rising every where; but though we have been throwing over them with all our skill for a quarter of an hour, yet not a single one will take; and I am afraid we shall return to breakfast without our prey.

* Vol. i. page 40.

"*Halieus.*—I will try; but I shall go to the other side, where I see a very large fish rising—There!—I have him at the very first throw—Land this fish, and put him into the well. Now, I have another; and I have no doubt I could take half-a-dozen in this very place, where you have been so long in fishing without success.

"*Physicus.*—You must have a different fly; or, have you some unguent or charm to tempt the fish?

"*Halieus.*—No such thing. If any of you will give me your rod and fly, I will answer for it I shall have the same success. I take your rod, Physicus—and lo! I have a fish!

"*Physicus.*—What can be the reason of this? It is perfectly inexplicable to me. Yet Poietes seems to throw as light as you do, and as well as he did yesterday.

"*Halieus.*—I am surprised that you, who are a philosopher, cannot discover the reason of this—Think a little.

"*All.*—We cannot.

"*Halieus.*—As you are my scholars, I believe I must teach you. The sun is bright, and you have been, naturally enough, fishing with your backs to the sun, which, not being very high, has thrown the shadows of your rods and yourselves upon the water, and you have alarmed the fish, wherever you have thrown a fly. You see, I have fished with my face towards the sun, and though inconvenienced by the light, have given no alarm. Follow my example, and you will soon have sport, as there is a breeze playing on the water.

"*Physicus.*—Your sagacity puts me in mind of

an anecdote which I remember to have heard respecting the late eloquent statesman, Charles James Fox; who, walking up Bond Street from one of the club-houses with an illustrious personage, laid him a wager, that he would see more cats than the Prince in his walk, and that he might take which side of the way he liked. When they got to the top, it was found, that Mr. Fox had seen thirteen cats, and the Prince not one. The Royal personage asked for an explanation of this apparent miracle, and Mr. Fox said, 'Your Royal Highness took, of course, the shady side of the way, as most agreeable; I knew that the sunny side would be left to me,—and cats always prefer the sunshine.'

"*Halieus.*—There! Poietes, by following my advice, you have immediately hooked a fish; and while you are catching a brace, I will tell you an anecdote, which as much relates to fly-fishing as that of Physicus, and affords an elucidation of a particular effect of light.

"A manufacturer of carmine, who was aware of the superiority of the French colour, went to Lyons for the purpose of improving his process, and bargained with the most celebrated manufacturer in that capital for the acquisition of his secret, for which he was to pay a thousand pounds. He was shown all the processes, and saw a beautiful colour produced, yet he found not the least difference in the French mode of fabrication and that which he had constantly adopted. He appealed to the manufacturer, and insisted that he must have concealed something. The manufacturer assured him that he had not, and invited him to see the process a second time. He minutely examined the water and the materials, which were the same as his own, and, very much surprised, said, 'I have lost my labour and my money, for the air of England does not permit us to make good carmine.'—'Stay,' says the Frenchman, 'do not deceive yourself; what kind of weather is it now?'—'A bright, sunny day,' said the Englishman.—'And such are the days,' said the Frenchman, 'on which I make my colour. Were I to attempt to manufacture it on a dark or cloudy day, my result would be the same as yours. Let me advise you, my friend, always to make carmine on bright and sunny days.'—'I will,' says the Englishman, 'but I fear I shall make very little in London.'

"*Poietes.*—Your anecdote is as much to the purpose as Physicus's; yet I am much obliged to you for the hint respecting the effect of shadow, for I have several times, in May and June, had to complain of too clear a sky, and wished, with Cotton, for

> 'A day with not too bright a beam;
> A warm, but not a scorching sun.'"

A very amusing and philosophical conversation on those natural phenomena, which have been vulgarly viewed as prophetic of dry or wet weather, may be well adduced as illustrative of that genius which, by the aid of a light of its own, imparts to the most trite objects all the charms of novelty.

"*Poietes.*—I hope we shall have another good day to-morrow, for the clouds are red in the west.

"*Physicus.*—I have no doubt of it, for the red has a tint of purple.

" *Halieus.*—Do you know why this tint portends fine weather?

" *Physicus.*—The air when dry, I believe, refracts more red, or heat-making rays; and as dry air is not perfectly transparent, they are again reflected in the horizon. I have generally observed a coppery or yellow sunset to foretell rain; but, as an indication of wet weather approaching, nothing is more certain than a halo round the moon, which is produced by the precipitated water; and the larger the circle, the nearer the clouds, and consequently the more ready to fall.

" *Halieus.*—I have often observed, that the old proverb is correct—

> ' A rainbow in the morning is the shepherd's warning;
> A rainbow at night is the shepherd's delight.'

—Can you explain this omen?

" *Physicus.*—A rainbow can only occur when the clouds containing or depositing the rain are opposite to the sun; and in the evening the rainbow is in the east, and in the morning in the west; and as our heavy rains in this climate are usually brought by the westerly wind, a rainbow in the west indicates that the bad weather is on the road, by the wind, to us; whereas the rainbow in the east proves, that the rain in these clouds is passing from us.

" *Poietes.*—I have often observed, that when the swallows fly high, fine weather is to be expected or continued; but when they fly low and close to the ground, rain is almost surely approaching. Can you account for this?

" *Halieus.*—Swallows follow the flies and gnats, and flies and gnats usually delight in warm strata of air; and as warm air is lighter, and usually moister than cold air, when the warm strata of air are high, there is less chance of moisture being thrown down from them by the mixture with cold air; but when the warm and moist air is close to the surface, it is almost certain, that as the cold air flows down into it, a deposition of water will take place.

" *Poietes.*—I have often seen sea-gulls assemble on the land, and have almost always observed, that very stormy and rainy weather was approaching. I conclude that these animals, sensible of a current of air approaching from the ocean, retire to the land to shelter themselves from the storm.

" *Ornither.*—No such thing. The storm is their element; and the little petrel enjoys the heaviest gale; because, living on the smaller sea insects, he is sure to find his food in the spray of a heavy wave. And you may see him flitting above the edge of the highest surge. I believe that the reason of the migration of sea-gulls and other sea birds to the land, is their security of finding food. They may be observed, at this time, feeding greedily on the earth worms and larvæ driven out of the ground by severe floods; and the fish on which they prey in fine weather in the sea, leave the surface when storms prevail, and go deeper. The search after food, as we agreed on a former occasion, is the principal cause why animals change their places. The different tribes of the wading birds always migrate when rain is about to take place; and I remember once in Italy having been long waiting, in the end of March, for the arrival of the double snipe in the

Campagna of Rome: a great flight appeared on the 3rd of April, and the day after heavy rain set in, which greatly interfered with my sport. The vulture, upon the same principle, follows armies; and I have no doubt, that the augury of the ancients was a good deal founded upon the observation of the instincts of birds. There are many superstitions of the vulgar owing to the same source. For anglers, in spring, it is always unlucky to see *single* magpies; but *two* may be always regarded as a favourable omen; and the reason is, that in cold and stormy weather, one magpie alone leaves the nest in search of food, the other remaining sitting upon the eggs or the young ones; but when two go out together, the weather is warm and mild, and thus favourable for fishing.

" *Poietes.*—The singular connexions of cause and effect, to which you have just referred, make superstition less to be wondered at, particularly amongst the vulgar; and when two facts, naturally unconnected, have been accidentally coincident, it is not singular that this coincidence should have been observed and registered, and that omens of the most absurd kind should be trusted in. In the West of England, half a century ago, a particular hollow noise on the sea-coast was referred to a spirit or goblin, called Bucca, and was supposed to foretell a shipwreck; the philosopher knows, that sound travels much faster than currents in the air—and the sound always foretold the approach of a very heavy storm, which seldom takes place on that wild and rocky coast, surrounded as it is by the Atlantic, without a shipwreck on some part of its extensive shores.*

"*Physicus.*—All the instances of omens you have mentioned are founded on reason; but how can you explain such absurdities as Friday being an unlucky day, the terror of spilling salt, or meeting an old woman? I knew a man of very high dignity, who was exceedingly moved by these omens, and who never went out shooting without a bittern's claw fastened to his button-hole by a riband—which he thought ensured him good luck.

"*Poietes.*—These, as well as the omens of death-watches, dreams, &c. are for the most part founded upon some accidental coincidences; but spilling of salt, on an uncommon occasion, may, as I have known it, arise from a disposition to apoplexy, shown by an incipient numbness in the hand, and may be a fatal symptom; and persons dispirited by bad omens sometimes prepare the way for evil fortune; for confidence in success is a great means of insuring it. The dream of Brutus, before the field of Philippi, probably produced a species of irresolution and despondency, which was the principal cause of his losing the battle; and I have heard, that the illustrious sportsman, to whom you

* Davy might also have adduced an equally striking superstition, in illustration of his subject, from the Cornish mines. The miners not unfrequently hear the echo of their own pickaxes, which they attribute to little fairies at work, and consider it as a happy omen. They say upon such occasions, that there will be good luck, as the Piskeys are at work. It is well known that the echo depends upon some cavity in the vicinity of the workmen,—and a cavity, or vogue, is always an indication of subterranean wealth.

referred just now, was always observed to shoot ill, because he shot carelessly, after one of his dispiriting omens.

" *Halieus.*—I have in life met with a few things which I found it impossible to explain, either by chance, coincidences, or by natural connexions; and I have known minds of a very superior class affected by them,—persons in the habit of reasoning deeply and profoundly.

" *Physicus.*—In my opinion, profound minds are the most likely to think lightly of the resources of human reason; it is the pert, superficial thinker who is generally strongest in every kind of unbelief. The deep philosopher sees chains of causes and effects so wonderfully and strangely linked together, that he is usually the last person to decide upon the impossibility of any two series of events being independent of each other; and in science, so many natural miracles, as it were, have been brought to light,—such as the fall of stones from meteors in the atmosphere, the disarming a thunder-cloud by a metallic point, the production of fire from ice by a metal white as silver, and referring certain laws of motion of the sea to the moon,—that the physical enquirer is seldom disposed to assert, confidently, on any abstruse subjects belonging to the order of natural things, and still less so on those relating to the more mysterious relations of moral events and intellectual natures."

Old Isaac Walton has amused us with a variety of absurd fables and superstitions: the author of Salmonia, on the other hand, touches, as with the spear of Ithuriel, the monsters and prodigies of the older writers, and they at once assume the forms of well-ascertained animals, or vegetables. The *sea snake* seen by American and Norwegian captains, appears as a company of porpoises, which in their gambols, by rising and sinking in lines, would give somewhat the appearance of the coils of a snake. The *Kraken*, or island fish, is reduced into an assemblage of *urticæ marinæ*, or sea blubbers. The *Mermaid*, into the long-haired seal;* and lastly, the celebrated Caithness Mermaid assumes the unpoetical form of a stout young traveller;—but this story is far too amusing to be dismissed with a passing notice.

"A worthy Baronet, remarkable for his benevolent views and active spirit, has propagated a story of this kind, and he seems to claim for his native country the honour of possessing this extraordinary animal; but the mermaid of Caithness was certainly a *gentleman*, who happened to be travelling on that wild shore, and who was seen bathing by some young ladies at so great a distance, that not only *genus* but gender was mistaken. I am acquainted with him, and have had the story from his own mouth. He is a young man, fond of geological pursuits, and one day in the middle of August, having fatigued and heated himself by climbing a rock to examine a particular appearance of granite,

* A pretended Mermaid was exhibited some time since in London, said to have been caught in the Chinese seas. It was soon discovered to have been manufactured by joining together the head and bust of two different apes to the lower part of a kipper salmon, which had the fleshy fin, and all the distinct characters of the *Salmo Salar*.

he gave his clothes to his Highland guide, who was taking care of his pony, and descended to the sea. The sun was just setting, and he amused himself for some time by swimming from rock to rock, and having unclipped hair and no cap, he sometimes threw aside his locks, and wrung the water from them on the rocks. He happened the year after to be at Harrowgate, and was sitting at table with two young ladies from Caithness, who were relating to a wondering audience the story of the mermaid they had seen, which had already been published in the newspapers: they described her, as she usually is described by poets, as a beautiful animal with remarkably fair skin and long green hair. The young gentleman took the liberty, as most of the rest of the company did, to put a few questions to the elder of the two ladies,—such as, on what day and precisely where this singular phenomenon had appeared. She had noted down not merely the day, but the hour and minute, and produced a map of the place. Our bather referred to his journal, and showed that a human animal was swimming in the very spot at that very time, who had some of the characters ascribed to the mermaid, but who laid no claim to others, particularly the green hair and fish's tail, but, being rather sallow in the face, was glad to have such testimony to the colour of his body beneath his garments."

With this story, I must conclude my review of "Salmonia,"—a work of considerable scientific and popular interest, and which cannot fail to become the favourite companion of the philosophical angler. The only production with which it can be at all compared is that of the "Complete Angler, by Izaac Walton." I agree with the critic who regards the two authors as pilgrims bound for the same shrine, resembling each other in their general habit—the scalloped hat, the dalmatique, and the knobbed and spiked staff—which equalize all who assume the character; yet, though alike in purpose, dress, and demeanour, the observant eye can doubtless discern an essential difference betwixt those devotees. The burgess does not make his approach to the shrine with the stately pace of a knight or a noble; the simple and uninformed rustic has not the contemplative step of the philosopher, or the quick glance of the poet. The palm of originality and of exquisite simplicity, which cannot perhaps be imitated with entire success, must remain with the common father of anglers—the patriarch Izaac; but it would be absurd to compare his work with the one written by the most distinguished philosopher of the nineteenth century, whose genius, like a sunbeam, illumined every recess which it penetrated, imparting to scarcely visible objects, definite forms and various colours.

If the advanced age of Walton was pleaded by himself as a sufficient reason for procuring "*a writ of ease*," the friends of Davy may surely claim at the hands of the critic an indulgent reception for a congenial work written in the hour of bodily lassitude and sickness. This benevolent feeling, however, did not penetrate every heart. A passage, which I shall presently quote, appears to have given great offence to the President of the Mechanics'

Institute, and to have been considered by him as the indication of a covert hostility to the spread of knowledge. The earth had scarcely closed upon the remains of the philosopher, when, in his anniversary speech,* the Autocrat of all the Mechanics, availing himself of this pretext, assailed his character with the charge of "conceit, pride, and arrogance."

The following is the passage in Salmonia, which provoked this angry and unjust philippic.

"I am sorry to say, I think the system carried too far in England. God forbid, that any useful light should be extinguished! let the persons who wish for education receive it; but it appears to me that, in the great cities in England, it is, as it were, forced upon the population; and that sciences, which the lower classes can only very superficially acquire, are presented to them; in consequence of which they often become idle and conceited, and above their usual laborious occupations. The unripe fruit of the tree of knowledge is, I believe, always bitter or sour; and scepticism and discontentment — sicknesses of the mind—are often the result of devouring it."

Methinks I hear the reader exclaim — "How little could Davy imagine that his prophetic words would have been so soon fulfilled!"—But I would seriously recommend to the President of the Me-

* See a Report of the President's speech, at the sixth anniversary of the Mechanics' Institute, as reported in all the journals of the day, December 5, 1829.

chanics' Institute, an anecdote which, if properly applied, cannot fail to be instructive.—When Diogenes, trampling with his dirty feet on the embroidered couch of Plato, cried out—"*Thus do I trample on the pride of Plato!*" the philosopher shook his head, and replied—"*Truly, but with more pride thou dost it, good Diogenes.*"

CHAPTER XV.

Sir H. Davy's paper on the Phenomena of Volcanoes.—His experiments on Vesuvius.—Theory of Volcanic action.—His reception abroad.—Anecdotes.—His last letter to Mr. Poole from Rome. —His paper on the Electricity of the Torpedo.—Consolations in Travel, or the Last Days of a Philosopher.—Analysis of the work.—Reflections suggested by its style and composition.—Davy and Wollaston compared.—His last illness.—Arrival at Geneva.—His Death.

A short time before Sir Humphry Davy quitted England, to which he was destined never to return, he communicated to the Royal Society a paper "On the Phenomena of Volcanoes;" which was read on the 20th of March 1828, and published in the Transactions of that year.

The object of this memoir was to collect and record the various observations and experiments which he had made on Vesuvius, during his several visits to that volcano. The appearances which it presented in 1814 and 1815 have been already noticed; it was in December 1819, and during the two succeeding months, that the mountain offered a favourable opportunity for making those experiments which form the principal subject of the present communication.

It was a point of great importance to determine

whether any combustion was going on at the moment the lava issued from the mountain; for this fact being once discovered, and the nature of the combustible matter ascertained, we should gain an immense step towards a just theory of the sources of volcanic action. For this purpose, he carefully examined both the lava and the elastic fluids with which it was accompanied. He was unable, however, to detect any thing like deflagration with nitre, which must have taken place had the smallest quantity of carbonaceous matter been present; nor could he, by exposing the ignited mass to portions of atmospheric air, discover that any appreciable quantity of oxygen had been absorbed. On immersing fused lava in water, no decomposition of that fluid followed, so that there could not have existed any quantity of the metallic bases of the alkalies or earths. Common salt, chloride of iron, the sulphates and muriates of potash, and soda, generally constituted the mass of solid products; while steam, muriatic acid fumes, and occasionally sulphurous acid vapours, formed the principal elastic matters disengaged.

He informs us it was on the 26th of January 1820, that he had the honour to accompany his Royal Highness the Prince of Denmark in an excursion to the mountain, on which occasion his friend, Cavalier Monticelli, was also present. At this time, the lava was seen nearly white hot through a chasm near the place where it flowed from the mountain; and yet, although he threw nitre upon it in large quantities through this chasm, there was no more increase of ignition than when the experiment was made on lava exposed to the free air. He observed that the appearance of the sublimations was very different from that which they had presented on former occasions; those near the aperture were coloured green and blue by salt of copper; but there was, as usual, a great quantity of muriate of iron. On the 5th, the sublimate of the lava was pure chloride of sodium; in the sublimate of January 6th, there were both sulphate of soda and indications of sulphate of potash; but in those which he collected during this last visit, the sulphate of soda was in much larger quantities, and there was much more of a salt of potash.

For nearly three months the craters, of which there were two, were in activity. The larger one threw up showers of ignited ashes and stones to a height apparently of from two hundred to two hundred and fifty feet; and from the smaller crater steam arose with great violence. Whenever the crater could be approached, it was found incrusted with saline matter: and the walk to the edge of the small crater, on the 6th of January, was through a mass of loose saline matter, principally common salt coloured by muriate of iron, in which the foot sunk to some depth. It was easy, even at a great distance, to distinguish between the steam disengaged by one of the craters, and the earthy matter thrown up by the other. The steam appeared white in the day, and formed perfectly white clouds, which reflected the morning and evening light of the purest tints of red and orange. The earthy matter always appeared as a black smoke, forming dark clouds, and in the night it was highly luminous at the moment of the explosion.

He concludes this paper on Volcanoes with some observations on the theory of their phenomena. "It appears," says he, "almost demonstrable, that none of the chemical causes anciently assigned for volcanic fires can be true. Amongst these, the combustion of mineral coal is one of the most current; but it seems wholly inadequate to account for the phenomena. However large the stratum of pit-coal, its combustion under the surface could never produce violent and excessive heat; for the production of carbonic acid gas, when there was no free circulation of air, must tend constantly to impede the process: and it is scarcely possible that carbonaceous matter, if such a cause existed, should not be found in the lava, and be disengaged with the saline or aqueous products from the bocca or craters. There are many instances in England of strata of mineral coal which have been long burning; but the results have been merely baked clay and schists, and it has produced no result similar to lava.

"If the idea of Lemery were correct, that the action of sulphur on iron may be a cause of volcanic fires, sulphate of iron ought to be the great product of the volcano; which is known not to be the case; and the heat produced by the action of sulphur on the common metals is quite inadequate to account for the appearances. When it is considered that volcanic fires occur and intermit with all the phenomena that indicate intense chemical action, it seems not unreasonable to refer them to chemical causes. But for phenomena upon such a scale, an immense mass of matter must be in activity, and the products of the volcano ought to give an idea of the nature of the substances primarily active. Now, what are these products? Mixtures of the earths in an oxidated and fused state, and intensely ignited; water and saline substances, such as might be furnished by the sea and air, altered in such a manner as might be expected from the formation of fixed oxidated matter. But it may be said, if the oxidation of the metals of the earths be the causes of the phenomena, some of these substances ought occasionally to be found in the lava, or the combustion ought to be increased at the moment the materials passed into the atmosphere. But the reply to this objection is, that it is evident that the changes which occasion volcanic fires take place in immense subterranean cavities; and that the access of air to the acting substances occurs long before they reach the exterior surface.

"There is no question but that the ground under the solfaterra is hollow; and there is scarcely any reason to doubt of a subterraneous communication between this crater and that of Vesuvius: whenever Vesuvius is in an active state, the solfaterra is comparatively tranquil. I examined the bocca of the solfaterra on the 21st of February 1820, two days before the activity of Vesuvius was at its height: the columns of steam which usually arise in large quantities when Vesuvius is tranquil, were now scarcely visible, and a piece of paper thrown into the aperture did not rise again; so that there was every reason to suppose the existence of a descending current of air. The subterraneous thunder heard at such great distances under Vesuvius is almost a demonstration of the existence of great cavities below filled with aëriform matter: and the same excavations which, in the active state of the volcano, throw out, during so great a length of time, immense volumes of steam, must, there is every reason to believe, in its quiet state, become filled with atmospheric air.*

"To what extent subterraneous cavities may exist, even in common rocks, is shown in the limestone caverns of Carniola, some of which contain many hundred thousand cubical feet of air; and in proportion as the depth of an excavation is greater, so is the air more fit for combustion.

"The same circumstances which would give alloys of the metals of the earths the power of producing volcanic phenomena, namely, their extreme facility of oxidation, must likewise prevent them from ever being found in a pure combustible state in the products of volcanic eruptions; for before they reach the external surface, they must not only be exposed to the air in the subterranean cavities, but be propelled by steam; which must possess, under the circumstances, at least the same facility of oxidating them as air. Assuming the hypothesis of the existence of such alloys of the metals of the earths as may burn into lava in the interior, the whole phenomena may be easily explained from the action of the water of the sea and air on those metals; nor is there any fact, or any of the circum-

* "Vesuvius is a mountain admirably fitted, from its form and situation, for experiments on the effect of its attraction on the pendulum: and it would be easy in this way to determine the problem of its cavities. On Etna the problem might be solved on a larger scale."

stances which I have mentioned in the preceding part of this paper, which cannot be easily explained according to that hypothesis. For almost all the volcanoes in the old world of considerable magnitude are near, or at no considerable distance from the sea: and if it be assumed that the first eruptions are produced by the action of sea-water upon the metals of the earths, and that considerable cavities are left by the oxidated metals thrown out as lava, the results of their action are such as might be anticipated; for, after the first eruptions, the oxidations which produce the subsequent ones may take place in the caverns below the surface; and when the sea is distant, as in the volcanoes of South America, they may be supplied with water from great subterranean lakes, as Humboldt states that some of them throw up quantities of fish.

"On the hypothesis of a chemical cause for volcanic fires, and reasoning from known facts, there appears to me no other adequate source than the oxidation of the metals which form the bases of the earths and alkalies; but it must not be denied, that considerations derived from thermometrical experiments on the temperature of mines and of sources of hot water, render it probable that the interior of the globe possesses a very high temperature: and the hypothesis of the nucleus of the globe being composed of fluid matter offers a still more simple solution of the phenomena of volcanic fires than that which has been just developed."

It must be admitted that the concluding sentence of this memoir is rather equivocal. He states that the metalloidal theory of volcanoes is most chemical, but that the hypothesis which assumes the high temperature of the interior of the globe is the most simple; but he leaves us in doubt as to his own belief upon the subject. In his "Last Days," however, we shall find that he offers a less reserved opinion upon this question.

With respect to Sir Humphry Davy's last journey to Rome, I have nothing of particular interest to relate. Universally known and respected, a member of almost every scientific society in Europe, there was not a part of the Continent in which he felt as a stranger in a foreign land. I might, in addition to the circumstances which have been already mentioned, relate several anecdotes in proof of the widely-extended popularity which his genius and discoveries had secured for him. The following striking incidents deserve particular notice.—Whilst sporting in Austria, he was assaulted by some peasants; and the outrage was no sooner made known to the Emperor, than he expressed his sorrow and indignation in the strongest language, and immediately directed that a party of troops should surround the district, and a most rigorous search be made for the culprits. The search was of course successful, and the "Carinthian boors" received merited chastisement.

For the following anecdote I am indebted to Lady Davy. Her Ladyship was travelling alone, on account of ill health, and upon arriving at Basle, she naturally felt a strong desire to visit its far-famed library; it so happened, however, that Sunday was the only day which afforded her this opportunity, and so strictly is the sabbath observed at that place, that she was at once informed that an admission to the library, under any circumstances, was altogether impossible. She nevertheless addressed a note to the librarian, stating to him her name, and the reasons for her unusual request. He immediately returned an answer, and appointed the hour of ten for her visit. Having shown her all that deserved inspection, he concluded his attentions by saying, "Madam, I have held the keys of this library for thirty years, during which period only three persons have been admitted to see its treasures on the Sunday; two of these were crowned heads, the third the wife of the most celebrated philosopher in Europe."

The following is the last letter which Davy ever wrote to his much-valued friend Mr. Poole.—

TO THOMAS POOLE, ESQ.

MY DEAR POOLE, Rome, Feb. 6, 1829.

I HAVE not written to you during my absence from England, because I had no satisfactory account of any marked progress towards health to give you, and the feelings of an invalid are painful enough for himself, and should, I think, never form a part of his correspondence; for they are not diminished by the conviction that they are felt by others. Would I were better! I would then write to you an agreeable letter from this glorious city; but I am here *wearing away* the winter; a ruin amongst ruins! I am anxious to hear from you,—very anxious, so pray write to me with this address, "Sir H. Davy, Inglese, posta restanti, Rovigo,

Italia." You know you must pay the postage to the frontier, otherwise the letters, like one a friend sent to me, will go back to you. Pray be so good as to be particular in the direction,—the "Inglese" is necessary. I hope you got a copy of my little trifle "*Salmonia*." I ordered copies to be sent to you, to Mr. W——, and to Mr. Baker: but as the course of letters in foreign countries is uncertain, I am not sure you received them; if not, you will have lost little; a *second edition* will soon be out, which will be in every respect more worthy of your perusal, being, I think, twice (not saying much for it) as entertaining and philosophical. I will take care by early orders that you have this book. I write and philosophize a good deal, and have nearly finished a work with a higher aim than the little book I speak of above, and which I shall dedicate to you. It contains the essence of my philosophical opinions, and some of my poetical reveries. It is, like the "Salmonia," an amusement of my sickness; but "*paulo majora canamus*." I sometimes think of the lines of Waller, and seem to feel their truth:

"The soul's dark cottage, batter'd and decay'd,
Lets in new lights through chinks that Time has made."

I have, notwithstanding my infirmities, attended to scientific objects whenever it was in my power, and I have sent the Royal Society a paper which they will publish, on the peculiar Electricity of the Torpedo, which I think bears remotely upon the functions of life. I attend a good deal to Natural History, and I think I have recognised in the Mediterranean a *new species of eel*, a sort of link between the conger and the muræna of the ancients. I have no doubt Mr. Baker is right about the distinction between the conger and the common eel. I am very anxious to hear what he thinks about *their generation*. Pray get from him a distinct opinion on this subject. I am at this moment getting the *eels in the markets* here dissected, and have found *ova* in plenty. Pray tell me particularly what Mr. Baker has done; this is a favourite subject with me, and you can give me no news so interesting. My dear friend, I shall never forget your kindness to me. You, with one other person, have given me the little happiness I have enjoyed since my severe visitation.

I fight against sickness and fate, believing I have still duties to perform, and that even my illness is connected in some way with my being made useful to my fellow-creatures. I have this conviction full on my mind, that intellectual beings spring from the same breath of Infinite Intelligence, and return to it again, but by different courses. Like rivers, born amidst the clouds of heaven, and lost in the deep and eternal ocean—some in youth, rapid and short lived torrents; some in manhood, powerful and copious rivers; and some in age, by a winding and slow course, half lost in their career, and making their exit through many sandy and shallow mouths. I hope to be at Rovigo about the first week in April. I travel slowly and with my own horses. If you will come and join me there, I can give you a place in a comfortable carriage, and can show you the most glorious country in Europe—Illyria and Styria, and take you to the French frontier before the beginning of autumn,—perhaps to England. If you can come, do so at once. I have two servants, and can accommodate you with every thing. I think of taking some baths before I return, in Upper Austria; but I write as if I were a strong man, when I am like a pendulum, as it were, swinging between death and life.

God bless you, my dear Poole.

Your grateful and affectionate friend,

H. DAVY.

Pray remember me to our friends at Stowey.

His paper on the Electricity of the Torpedo, to which he alludes in the foregoing letter, appears to have been written shortly after he had finished his "Salmonia," as it is dated from Lubiana, Illyria, on the 24th of October, and it was read before the Royal Society on the 20th of November 1828, and published in the first part of the Transactions for 1829. It will be remembered, that this subject had long engaged his attention; and he expresses his surprise that the electricity of living animals should not have been an object of greater attention, both on account of its physiological importance, and its general relation to the science of electro-chemistry.

When Volta discovered his wonderful pile, he imagined he had made a perfect resemblance of the organ of the Gymnotus and Torpedo; and Davy observes, that whoever has felt the shocks of the natural and artificial instruments must have been convinced, as far as sensation is concerned, of their strict analogy.

After the discovery of the *chemical* power of the Voltaic instrument, he was naturally desirous of ascertaining whether this property was possessed by the electrical organs of living animals; for which purpose, he instituted various experiments, but he could not discover that such was the fact. Upon mentioning his researches to Signor Volta, with whom he passed some time in the summer of 1815, the Italian philosopher showed him a peculiar form of his instrument, which appeared to fulfil the conditions of the organs of the torpedo; *viz.* a pile, of which the fluid substance was a very imperfect conductor, such as honey or a strong saccharine extract, which required a certain time to become charged, and which did not decompose water, though it communicated weak shocks.

The discovery by Oersted of the effects of Voltaic electricity on the magnetic needle induced Davy to examine whether the electricity of living animals possessed a similar power. Having, after some trouble, procured two lively and recently caught Torpedoes, he passed the shocks from the largest of these animals a number of times through the circuit of an extremely delicate magnetic electrometer, but, although every precaution was used, not the slightest deviation of, or effect on, the needle could be perceived.

"These negative results," says he, "may be explained by supposing that the motion of the electricity in the torpedinal organ is in no measurable time, and that a current of some continuance is necessary to produce the deviation of the magnetic needle; and I found that the magnetic electrometer was equally insensible to the weak discharge of a Leyden jar as to that of the torpedinal organ;

though whenever there was a continuous current from the smallest surfaces in Voltaic combinations of the weakest power, but in which some chemical action was going on, it was instantly and powerfully affected. Two series of zinc and silver, and paper moistened in salt and water, caused the permanent deviation of the needle several degrees, though the plates of zinc were only one-sixth of an inch in diameter.

"It would be desirable to pursue these enquiries with the electricity of the Gymnotus, which is so much more powerful than that of the Torpedo; but if they are now to be reasoned upon, they seem to show a stronger analogy between common and animal electricity, than between Voltaic and animal electricity; it is however, I think, more probable that animal electricity will be found of a distinctive and peculiar kind.

"Common electricity is excited upon non-conductors, and is readily carried off by conductors and imperfect conductors. Voltaic electricity is excited upon combinations of perfect and imperfect conductors, and is only transmitted by perfect conductors, or imperfect conductors of the best kind.

"Magnetism, if it be a form of electricity, belongs only to perfect conductors; and, in its modifications, to a peculiar class of them.

"The animal electricity resides only in the imperfect conductors forming the organs of living animals, and its object in the economy of nature is to act on living animals.

"Distinctions might be established in pursuing the various modifications or properties of electricity in these different forms; but it is scarcely possible to avoid being struck by another relation of this subject. The torpedinal organ depends for its powers upon the will of the animal. John Hunter has shown how copiously it is furnished with nerves. In examining the columnar structure of the organ of the Torpedo, I have never been able to discover arrangements of different conductors similar to those in galvanic combinations, and it seems not improbable that the shock depends upon some property developed by the action of the nerves.

"To attempt to reason upon any phenomena of this kind as dependent upon a specific fluid would be wholly vain. Little as we know of the nature of electrical action, we are still more ignorant of the nature of the functions of the nerves. There seems, however, a gleam of light worth pursuing in the peculiarities of animal electricity,—its connexion with so large a nervous system, its dependence upon the will of the animal, and the instantaneous nature of its transfer, which may lead, when pursued by adequate enquirers, to results important for physiology."

He concludes this paper by expressing his fear that the weak state of his health will prevent him from following the subject with the attention it seems to deserve; and he therefore communicates these imperfect trials to the Royal Society, in the hope that they may lead to more extensive and profound researches.

We come now to the consideration of the last production of his genius—"Consolations in Travel, or the Last Days of a Philosopher:" A work which, he informs us in the preface, was composed immediately after Salmonia, under the same unfavourable and painful circumstances, and at a period when his constitution suffered from new attacks. From this exercise of the mind, he tells us, that he derived some pleasure and some consolation, when most other sources of consolation and pleasure were closed to him; and he ventures to hope that those hours of sickness may be not altogether unprofitable to persons in perfect health. His brother, Dr. Davy, who edited the work after the decease of Sir Humphry, informs us that it was concluded at the very moment of the invasion of the author's last illness, and that, had his life been prolonged, it is probable some additions and some changes would have been made.

"The characters of the persons of the dialogue," continues the Editor, "were intended to be ideal, at least in great part;—such they should be considered by the reader; and it is to be hoped, that the incidents introduced, as well as the persons, will be viewed only as subordinate and subservient to the sentiments and the doctrines. The dedication, it may be specially noticed, is the Author's own, and in the very words dictated by himself at a time when he had lost the power of writing, except with extreme difficulty, owing to the paralytic attack, although he retained in a very remarkable manner all his mental faculties unimpaired and unclouded." The words of the Dedication are "To Thomas Poole, Esq. of Nether Stowey; in remembrance of thirty years of continued and faithful friendship."

This is a most extraordinary and interesting work: extraordinary, not only from the wild strength of its fancy, and the extravagance of its conceptions, but from the bright light of scientific truth which is constantly shining through its metaphorical tissue, and irradiating its most shadowy imaginings. It may be compared to the tree of the lower regions in the Æneid, to every leaf of which was attached a dream; and yet, however wildly his fancy may dream, his philosophy never sleeps; and in his exit from the land of phantoms, the author can in no instance be accused of having mistaken the gate of ivory for that of horn. To the biographer, the work is of the highest interest and value, by confirming, in a remarkable manner, the opinion so frequently expressed in the course of these memoirs, with respect to the diversified talents of Sir Humphry Davy; and above all, by elucidating that rare combination of imagination with judgment, which imparted to his genius its more striking peculiarities.

The work consists of six Dialogues:—1. THE VISION; 2. DISCUSSIONS CONNECTED WITH THE VISION IN THE COLOSÆUM; 3. THE UNKNOWN; 4. THE PROTEUS, OR IMMORTALITY; 5. THE CHEMICAL PHILOSOPHER; and 6. POLA, OR TIME.

The interlocutors of the first dialogue are two intellectual Englishmen, one of whom the author calls *Ambrosio;* a man of highly cultivated taste, great classical erudition, and minute historical knowledge: a Catholic in religion, but so liberal in his sentiments, that in another age he might have been secretary to Ganganelli. The other friend, whom he calls *Onuphrio,* was a man of a very different character: belonging to the English aristocracy, he had some of the prejudices usually attached to birth and rank; but his manners were gentle, his temper good, and his disposition amiable. Having been partly educated at a northern university in Britain, he had adopted views in religion which went even beyond toleration, and which might be regarded as entering the verge of scepticism. For a patrician, he was very liberal in his political views. His imagination was poetical and discursive, his taste good, and his tact extremely fine,—so exquisite, indeed, that it sometimes approached to morbid sensibility, and disgusted him with slight defects, and made him keenly sensible of small perfections to which common minds have been indifferent.

The author, with these his two friends, makes an excursion to the Colosæum, and the conversation, which a view of those magnificent ruins produced, together with the account of a dream, or vision, which occurred to him while left alone amidst these mouldering monuments, forms the subject-matter of the first dialogue. It is impossible for any person of the least imagination to contemplate this decay of former magnificence without strong emotion; but the direction and tone of such feeling will be necessarily modified by the qualities of the mind in which it is excited; and the author has therefore very properly assigned to each of the *dramatis personæ,* such opinions as might best correspond with his character and temperament.

They are all represented as being struck with the transiency of human monuments; but *Ambrosio* views with triumph the sanctifying influence of a few crosses planted around the ruins, in arresting the farther decay of the pile. "Without the influence of Christianity," he exclaims, "these majestic ruins would have been dispersed or levelled to the dust. Plundered of their lead and iron by the barbarians, Goths and Vandals, and robbed even of their stones by Roman princes—the Barberini, they owe what remains of their relics to the sanctifying influence of that faith which has preserved for the world all that was worth preserving, not merely arts and literature, but likewise that which constitutes the progressive nature of intellect, and the institutions which afford to us happiness in this world, and hopes of a blessed immortality in the next." And he continues,—"What a contrast the present application of this building, connected with holy feelings and exalted hopes, is to that of the ancient one, when it was used for exhibiting to the Roman people the destruction of men by wild beasts, or of men more savage than wild beasts by each other, to gratify a horrible appetite for cruelty, founded upon a still more detestable lust, that of universal domination! And who would have supposed, in the time of Titus, that a faith, despised in its insignificant origin, and persecuted from the supposed obscurity of its founder and its principles, should have reared a dome to the memory of one of its humblest teachers, more glorious than was ever framed for Jupiter or Apollo in the ancient world, and have preserved even the ruins of the temples of the Pagan deities, and have burst forth in splendour and majesty, consecrating truth amidst the shrines of error, employing the idols of the Roman superstition for the most holy purposes, and rising a bright and constant light amidst the dark and starless night which followed the destruction of the Roman empire!"

It was not to be expected that *Onuphrio*, whose views are represented as verging upon scepticism, should have tacitly coincided in these opinions of *Ambrosio.* He admits, indeed, that some little of the perfect state in which these ruins exist may have been owing to the causes just described; but these causes, he maintains, have only lately begun to operate, and the mischief was done before Christianity was established at Rome. "Feeling differently on these subjects," says he, "I admire this venerable ruin rather as the record of the destruction of the power of the greatest people that ever existed, than as a proof of the triumph of Christianity; and I am carried forward, in melancholy anticipation, to the period when even the magnificent dome of St. Peter's will be in a similar state to that in which the Colosæum now is, and when its ruins may be preserved by the sanctifying influence of some new and unknown faith; when perhaps the statue of Jupiter, which at present receives the kiss of the devotee, as the image of St. Peter, may be employed for another holy use, as the personification of a future saint or divinity; and when the monuments of the Papal magnificence shall be mixed with the same dust as that which now covers the tombs of the Cæsars.

"Such, I am sorry to say, is the general history of all the works and institutions belonging to humanity. They rise, flourish, and then decay and fall; and the period of their decline is generally proportional to that of their elevation. In ancient Thebes or Memphis, the peculiar genius of the people has left us monuments from which we can judge of their arts, though we cannot understand the nature of their superstitions. Of Babylon and of Troy, the remains are almost extinct; and what we know of those famous cities is almost entirely derived from literary records. Ancient Greece and Rome we view in the few remains of their monuments; and the time will arrive when modern Rome shall be what ancient Rome now is; and ancient Rome and Athens will be what Tyre or Carthage now are, known only by coloured dust in the desert, or coloured sand, containing the fragments of bricks, or glass, washed up by the wave of a stormy sea."

For this desponding view of passing events, *Onuphrio* finds consolation in the evidences of revealed religion. In the origin, progress, elevation, decline, and fall of the empires of antiquity, he sees proofs that they were intended for a definite end in the scheme of human redemption; and he finds prophecies which have been amply verified. He regards the foundation or the ruin of a kingdom, which appears in civil history so great an event, as comparatively of small moment in the history of man and in his religious institutions. He considers the establishment of the worship of one God amongst a despised and contemned people, as the most important circumstance in the history of the early world. He regards the Christian dispensation as naturally arising out of the Jewish, and the doctrines of the Pagan nations all preparatory to the triumph and final establishment of a creed fitted for the most enlightened state of the human mind, and equally adapted to every climate and every people."

We cannot but regard these passages with great interest, as indicating the train of thought which must have occupied the mind of their author, and as proving that, in his latter days, he not only studied the doctrines of Christianity, but derived the greatest consolation from its tenets.

After some farther conversation, *Onuphrio* and *Ambrosio* leave their friend the author to pursue his meditations amidst the solitude of the ruins.

Seated in the moonshine on one of the steps leading to the seats supposed to have been occupied by the patricians in the Colosæum at the time of the public games, the train of ideas in which he had before indulged continued to flow with a vividness and force increased by the stillness and solitude of the scene, and by the full moon, which, he observes, has always a peculiar effect on these moods of feeling in his mind, giving to them a wildness and a kind of indefinite sensation, such as he supposes belong at all times to the true poetical temperament.

"It must be so," thought he, "no new city will rise again out of the double ruins of this; no new empire will be founded upon these colossal remains of that of the old Romans. The world, like the individual, flourishes in youth, rises to strength in manhood, falls into decay in age; and the ruins of an empire are like the decrepid frame of an individual, except that they have some tints of beauty, which nature bestows upon them. The sun of civilization arose in the East, advanced towards the

West, and is now at its meridian; in a few centuries more, it will probably be seen sinking below the horizon, even in the new world; and there will be left darkness only where there is a bright light, deserts of sand where there were populous cities, and stagnant morasses where the green meadow, or the bright corn-field once appeared. Time," he exclaimed, "which purifies, and, as it were, sanctifies the mind, destroys and brings into utter decay the body; and even in nature its influence seems always degrading. She is represented by the poet as eternal in her youth; but amongst these ruins she appears to me eternal in her age, and here, no traces of renovation appear in the ancient of days."

He had scarcely concluded this ideal sentence, when his reverie became deeper, and his imagination called up a spirit, who, having rebuked him for his ignorance and presumption, undeceives him in his views of the history of the world, by unfolding to him in a vision the progress of man from a state of barbarity to that of high civilization. He is first shown a country covered with forests and marshes; wild animals were grazing in large savannahs, and carnivorous beasts, such as lions and tigers, occasionally disturbing and destroying them. Man appeared as a naked savage, feeding upon wild fruits, or devouring shell-fish, or fighting with clubs for the remains of a whale which had been thrown upon the shore. His habitation was a cave in the earth —"See the birth of Time!" exclaimed the Genius; "look at man in his newly-created state, full of youth and vigour. Do you see aught in this state to admire or envy?"

In the next scene, a country opened upon his view, which appeared partly wild and partly cultivated; and men were seen covered with the skins of animals, and driving cattle to enclosed pastures; others were reaping and collecting corn, and others again were making it into bread. Cottages appeared furnished with many of the conveniences of life. The Genius now said, "Look at these groups of men who are escaped from the state of infancy; they owe their improvement to a few superior minds still amongst them. That aged man whom you see with a crowd around him taught them to build cottages; from that other they learnt to domesticate cattle; from others, to collect and sow corn and seeds of fruit. And these arts will never be lost; another generation will see them more perfect. You shall be shown other visions of the passages of time; but as you are carried along the stream which flows from the period of creation to the present moment, I shall only arrest your transit to make you observe some circumstances which will demonstrate the truths I wish you to know." He then proceeds to describe in succession the different scenes as they appeared before him, and to relate the observations by which his genius, or intellectual guide, accompanied him.

A great extent of cultivated plains, large cities on the sea-shore, palaces, forums, and temples, were displayed before him. He saw men associated in groups, mounted on horses, and performing military exercises; galleys moved by oars on the ocean, roads intersecting the country covered with travellers, and containing carriages moved by men or horses. The Genius now said, " You see the early state of civilization of man: the cottages of the last race you beheld have become improved into stately dwellings, palaces, and temples, in which use is combined with ornament. The few men to whom, as I said before, the foundations of these improvements were owing, have had divine honours paid to their memory. But look at the instruments belonging to this generation, and you will find they were only of brass. You see men who are talking to crowds around them, and others who are apparently amusing listening groups by a kind of song or recitation; these are the earliest bards and orators; but all their signs of thought are oral, for written language does not yet exist."

The Genius next presented to him a scene of varied business and imagery. He saw a man who bore in his hands the same instruments as our modern smiths, presenting a vase, which appeared to be made of iron, amidst the acclamations of an assembled multitude; and he saw in the same place men who carried rolls of papyrus in their hands, and wrote upon them with reeds containing ink, made from the soot of wood mixed with a solution of glue.—" See," the Genius said, " an immense change produced in the condition of society by the two arts of which you now see the origin; the one, that of rendering iron malleable, which is owing to a single individual, an obscure Greek; the other, that of making thought permanent in written characters, —an art which has gradually arisen from the hieroglyphics which you may observe on yonder pyramids. You will now see human life more replete with power and activity."

In the scenes that succeeded, he saw bronze instruments thrown away; malleable iron converted into hard steel, and applied to a thousand purposes of civilized life; bands of men traversing the sea, founding colonies, building cities, and, wherever they established themselves, carrying with them their peculiar arts. He saw the Roman world succeeded by cities filled with an idle and luxurious population, and the farms which had been cultivated by warriors, who left the plough to take the command of armies, now in the hands of slaves; and the militia of free men supplanted by bands of mercenaries, who sold the Empire to the highest bidder. He saw immense masses of warriors collecting in the North and East, carrying with them no other proofs of cultivation but their horses and steel arms. He saw these savages every where plundering cities and destroying the monuments of arts and literature. Ruin, desolation, and darkness were before him, and he closed his eyes to avoid the melancholy scene. "See," said the Genius, "the termination of a power believed by its founders invincible, and intended to be eternal. But you will find, though the glory and greatness belonging to its military genius have passed away, yet those belonging to the arts and institutions by which it adorned and dignified life, will again arise in another state of society."

Upon again opening his eyes, he saw Italy recovering from her desolation, towns arising with governments almost upon the model of ancient Athens and Rome, and these different small states rivals in arts and arms;—he saw the remains of libraries, which had been preserved in monasteries and churches by a holy influence, which even the Goth and Vandal respected, again opened to the people;—he saw Rome rising from her ashes, the fragments of statues found amidst the ruins of her palaces and imperial villas, becoming the models for the regeneration of art;—he saw magnificent temples raised in this city, become the metropolis of a new and Christian world, and ornamented with the most brilliant master-pieces of the arts of design.—"Now," the Genius said, "society has taken its modern and permanent aspect. Consider for a moment its relations to letters and to arms, as contrasted with those of the ancient world." He looked, and he saw that, in the place of the rolls of papyrus, libraries were now filled with books. "Behold," the Genius said, "THE PRINTING PRESS! By the invention of Faust, the productions of genius are, as it were, made imperishable, capable of indefinite multiplication, and rendered an unalienable heritage of the human mind. By this art, apparently so humble, the progress of society is secured, and man is spared the humiliation of witnessing again scenes like those which followed the destruction of the Roman Empire. Now look to the warriors of modern times; you see the spear, the javelin, and the cuirass are changed for the musket and the light artillery. The German monk who discovered gunpowder did not meanly affect the destinies of mankind; wars are become less bloody by becoming less personal; mere brutal strength is rendered of comparatively little avail; all the resources of civilization are required to move a large army; wealth, ingenuity, and perseverance become the principal elements of success; civilized man is rendered in consequence infinitely superior to the savage, and gunpowder gives permanence to his triumph, and secures the cultivated nations from being ever again overrun by the inroads of millions of barbarians."*

The Genius then directs his attention to scenes in which are displayed the triumphs of modern science; such as the steam-engine, and the thousand resources furnished by the chemical and mechanical arts; and she concludes by endeavouring to impress upon him the conviction, "That the results of intellectual labour, or of scientific genius, are permanent, and incapable of being lost. Monarchs change their plans, governments their objects, a fleet or an army effect their purpose and then pass away; but a piece of steel touched by the magnet preserves its character for ever, and secures to man the dominion of the trackless ocean. A new period of society may

* This is a question which Gibbon has very eloquently discussed ("*General Observations on the Fall of the Roman Empire in the West*," vol. vi.) "Cannon and fortifications now form an impregnable barrier against the Tartar horse; and Europe is secure from any future irruption of barbarians; since, before they can conquer, they must cease to be barbarous." What an extraordinary illustration does this principle find in the history of our possessions in India, where, to speak in round numbers, thirty thousand Europeans keep no less than one hundred million of natives in subjection!

send armies from the shores of the Baltic to those of the Euxine, and the empire of the followers of Mahomet may be broken in pieces by a Northern people, and the dominion of the Britons in Asia may share the fate of Tamerlane or Zengis-khan; but the steam-boat which ascends the Delaware, or the St. Lawrence, will be continued to be used, and will carry the civilization of an improved people into the deserts of North America, and into the wilds of Canada. In the common history of the world, as compiled by authors in general, almost all the great changes of nations are confounded with changes in their dynasties, and events are usually referred either to sovereigns, chiefs, heroes, or their armies, which do, in fact, originate from entirely different causes, either of an intellectual or moral nature."

Having instructed him in the history of man as an inhabitant of the earth, the Genius proceeds to reveal to him the mysteries of spiritual natures, in which the author evidently shows his attachment to the belief that our intellectual essence is destined hereafter to enjoy a higher and better state of planetary existence,* drinking intellectual light from a purer source, and approaching nearer to the Infinite and Divine mind. I shall not attempt to follow him and his Genius to the verge of the solar system,

* Under the article 'Sensation,' in the Philosophical Dictionary, we find Voltaire indulging in a similar speculation. "It may be, that in other globes the inhabitants possess sensations of which we can form no idea. It is possible that the number of our senses augments from globe to globe, and that an existence with innumerable and perfect senses will be the final attainment of all being."

witnessing in his career the inhabitants of planets and comets. We may upon this occasion truly apply to the author the words of Lucretius—

> "Processit longe flammantia mœnia mundi."
>
> "His vigorous and active mind was hurl'd
> Beyond the flaming limits of the world."—CREECH.

In the former part of the dialogue, his poetical coruscations appeared only as brilliant sparks thrown off by the rapidity of the machinery which he worked for a useful end and for a definite purpose; his vivid imagination may now be compared to a display of fire-works, which dazzle and confound without enlightening the senses, and leave the spectator in still more profound darkness.

His SECOND DIALOGUE, entitled "Discussions connected with the Vision in the Colosæum," may be considered as a commentary upon the views he had unfolded; and a more appropriate spot, perhaps, could not have been selected for a conversation upon the progress of civilization, than the summit of Vesuvius, from which, to adopt the language of *Ambrosio*, "We see not only the power and activity of man as existing at present, and of which the highest example may be represented by the steam-boat departing from Palermo, but we may likewise view scenes which carry us into the very bosom of antiquity, and as it were make us live with the generations of past ages."

The author, who assumes throughout this dialogue the name of *Philalethes*, after having been duly rallied by his friends on the subject of his vision, thus expresses himself:—"I will acknowledge that the vision in the Colosæum is a fiction;

but the most important parts of it really occurred to me in sleep, particularly that in which I seemed to leave the earth and launch into the infinity of space, under the guidance of a tutelary genius. And the origin and progress of civil society form likewise parts of another dream which I had many years ago; and it was in the reverie which happened when you quitted me in the Colosæum, that I wove all these thoughts together, and gave them the form in which I narrated them to you.—I do not say that they are strictly to be considered as an accurate representation of my waking thoughts; for I am not quite convinced that dreams are always the representations of the state of the mind, modified by organic diseases or by associations. There are certainly no absolutely new ideas produced in sleep; yet I have had more than one instance, in the course of my life, of most extraordinary combinations occurring in this state, which have had considerable influence on my feelings, my imagination, and my health."

Philalethes now relates a fact to which his preceding observation more immediately referred; he anticipates unbelief,—but he declares that he mentions nothing but a simple fact.

"Almost a quarter of a century ago, I contracted that terrible form of typus fever known by the name of jail fever,—I may say, not from any imprudence of my own, but whilst engaged in putting in execution a plan for ventilating one of the great prisons of the metropolis.* My illness was severe and dan-

* See page 287, vol. i. for an account of this event.

gerous; as long as the fever continued, my dreams and deliriums were most painful and oppressive; but when weakness consequent to exhaustion came on, and when the probability of death seemed to my physicians greater than that of life, there was an entire change in all my ideal combinations. I remained in an apparently senseless or lethargic state, but, in fact, my mind was peculiarly active; there was always before me the form of a beautiful woman, with whom I was engaged in the most interesting and intellectual conversation."

Ambrosio and *Onuphrio* very naturally suggest that this could have been no other than the image of some favourite maiden which had haunted his imagination; but *Philalethes* rejects with indignation such an explanation of the vision. "I will not," he exclaims, "allow you to treat me with ridicule on this point, till you have heard the second part of my tale. Ten years after I had recovered from the fever, and when I had almost lost the recollection of the vision, it was recalled to my memory by a very blooming and graceful maiden fourteen or fifteen years old, that I accidentally met during my travels in Illyria; but I cannot say that the impression made upon my mind by this female was very strong. Now comes the extraordinary part of the narrative: ten years after,—twenty years after my first illness, at a time when I was exceedingly weak from a severe and dangerous malady, which for many years threatened my life, and when my mind was almost in a desponding state, being in a course of travels ordered by my medical advisers, I again met the person who was the representative of my visionary female; and to her kindness and care, I believe, I owe what remains to me of existence. My despondency gradually disappeared, and though my health still continued weak, life began to possess charms for me which I had thought were for ever gone; and I could not help identifying the living angel with the vision which appeared as my guardian genius during the illness of my youth."

The reader will probably agree with *Onuphrio*, in seeing in this history nothing beyond the influence of an imagination excited by disease.

The discourse now turns upon that part of the vision in the Colosæum in which was exhibited the early state of man, after his first creation, and which *Ambrosio* considers as not only incompatible with revelation, but likewise with reason and every thing that we know respecting the history or traditions of the early nations of antiquity.

I shall merely state the objection which *Ambrosio* offers. I must then refer the reader to the work itself for an account of the discussion it provoked.

"*Ambrosio.*—You consider man, in his early state, a savage like those who now inhabit New Holland, or New Zealand, acquiring, by the little use that they make of a feeble reason, the power of supporting and extending life. Now, I contend that, if man had been so created, he must inevitably have been destroyed by the elements, or devoured by savage beasts, so infinitely his superiors in physical force."

During the discussion, an opinion is advanced by *Ambrosio*, so singular, that I must be allowed to quote it. "I consider," says he, "all the miraculous parts of our religion as effected by changes in the sensations or ideas of the human mind, and not by physical changes in the order of nature! To Infinite Wisdom and Power, a change in the intellectual state of the human being may be the result of a momentary will, and the mere act of faith may produce the change. How great the powers of imagination are, even in ordinary life, is shown by many striking facts, and nothing seems impossible to this imagination when acted upon by Divine influence."

This is surely a most extraordinary line of argument for the apologist of the Christian faith, and of the miracles by which it is supported.

In the THIRD DIALOGUE, called the Unknown, the author and his friends, *Ambrosio* and *Onuphrio*, make an excursion to the remains of the temples of Pæstum. "Were my existence to be prolonged through ten centuries," exclaims the author, "I think I could never forget the pleasure I received on that delicious spot." In contemplating beautiful scenery, much of its interest depends upon the feelings and associations of the moment; and the author was upon this occasion evidently in that poetical frame of mind which sheds a magic light over every landscape, and converts the most ordinary objects into emblems of morality: in the admixture of the olive and the cypress tree, he saw a connection, to memorialize, as it were, how near each other are life and death, joy and sorrow; while the music of the birds, and, above all, the cooing of the turtle-doves, by overpowering the murmuring of the waves and the whistling of the winds, served but to show him that, in the strife of nature, the voice of love is predominant.

With their hearts touched by the scene they had witnessed, the travellers descended to the ruins, and began to examine those wonderful remains which have outlived even the name of the people by whom they were raised. While engaged in measuring the Doric columns in the interior of the Temple of Neptune, a stranger, remarkable both in dress and appearance, was observed to be writing in a memorandum book; the author immediately addresses him, and becoming mutually pleased with each other, they enter into a conversation of high scientific interest.

The sentiments delivered by the "UNKNOWN," for by this title is the philosopher designated, notwithstanding their dramatic dress, are evidently to be received as the bequest of the latest scientific opinions of Sir H. Davy upon several important subjects, and must consequently command our respect and consideration.

To a question relative to the nature of the masses of travertine, of which the ruins consisted, the Unknown replied, that they were certainly produced by deposition from water; and he rather believed, that a lake in the immediate neighbourhood of the city furnished the quarry. The party are then described as visiting this spot.

"There was something peculiarly melancholy in the character of this water; all the herbs around it were grey, as if incrusted with marble; a few buffaloes were slaking their thirst in it, which ran wildly away at our approach, and appeared to retire into a rocky excavation or quarry at the end of the lake. 'There,' said the stranger, 'is what I believe to be the source of those large and durable stones which you see in the plain before you. This water rapidly deposits calcareous matter, and even, if you throw a stick into it, a few hours is sufficient to give it a coating of this substance. Whichever way you turn your eyes, you see masses of this recently produced marble, the consequence of the overflowing of the lake during the winter floods.'

* * * * *

"This water is like many, I may say most, of the sources which rise at the foot of the Apennines; it holds carbonic acid in solution, which has dissolved a portion of the calcareous matter of the rock through which it has passed:—this carbonic acid is dissipated in the atmosphere, and the marble, slowly thrown down, assumes a crystalline form, and produces coherent stones. The lake before us is not particularly rich in the quantity of calcareous matter, for, as I have found by experience, a pint of it does not afford more than five or six grains; but the quantity of fluid and the length of time are sufficient to account for the immense quantities of tufa and rock which, in the course of ages, have accumulated in this situation.

* * * * *

"It can, I think, be scarcely doubted that there is a source of volcanic fire at no great distance from the surface, in the whole of southern Italy; and, this fire acting upon the calcareous rocks of which the Apennines are composed, must constantly detach from them carbonic acid, which rising to the sources of the springs, deposited from the waters of the atmosphere, must give them their impregnation, and enable them to dissolve calcareous matter. I need not dwell upon Ætna, Vesuvius, or the Lipari Islands, to prove that volcanic fires are still in existence; and there can be no doubt that, in earlier periods, almost the whole of Italy was ravaged by them; even Rome itself, the eternal city, rests upon the craters of extinct volcanoes; and I imagine that the traditional and fabulous record of the destruction made by the conflagration of Phaeton, in the chariot of the Sun, and his falling into the Po, had reference to a great and tremendous igneous volcanic eruption which extended over Italy, and ceased only near the Po, at the foot of the Alps. Be this as it may, the sources of carbonic acid are numerous, not merely in the Neapolitan but likewise in the Roman and Tuscan states. The most magnificent waterfall in Europe, that of the Velino near Terni, is partly fed by a stream containing calcareous matter dissolved by carbonic acid, and it deposits marble, which crystallizes even in the midst of its thundering descent and foam, in the bed in which it falls.

"There is a lake in Latium, a few yards above the Lacus Albula, where the ancient Romans erected their baths, which sends down a considerable stream of tepid water to the larger lake; but this water is less strongly impregnated with carbonic acid; the largest lake is actually a saturated solution of this gas, which escapes from it in such quantities in some parts of its surface, that it has the appearance of being actually in ebullition. Its temperature I ascertained to be, in the winter, in the warmest parts, above 80 degrees of Fahrenheit, and as it appears to be pretty constant, it must be supplied with heat from a subterraneous source, being nearly twenty degrees above the mean temperature of the atmosphere. Kircher has detailed, in his *Mundus Subterraneus*, various wonders respecting this lake, most of which are unfounded; such as, that it is unfathomable,—that it has at the bottom the heat of boiling water, and that floating islands rise from the gulf which emits it. It must certainly be very difficult, or even impossible, to fathom a source which rises with so much violence from a subterraneous excavation; and at a time when chemistry had made small progress, it was easy to mistake the disengagement of carbonic acid for an actual ebullition. The floating islands are real; but neither the Jesuit, nor any of the writers who have since described this lake, had a correct idea of their origin, which is exceedingly curious. The high temperature of this water, and the quantity of carbonic acid that it contains, render it peculiarly fitted to afford a pabulum or nourishment to vegetable life; the banks of travertine are every where covered with reeds, lichens, confervæ, and various kinds of aquatic vegetables; and at the same time that the process of vegetable life is going on, the crystallizations of the calcareous matter, which is every where deposited in consequence of the escape of carbonic acid, likewise proceed, giving a constant milkiness to what from its tint would otherwise be a blue fluid. So rapid is the vegetation, owing to the decomposition of the carbonic acid, that even in winter, masses of confervæ and lichens, mixed with deposited travertine, are constantly detached by the current of water from the bank, and float down the stream, which being a considerable river, is never without many of these small islands on its surface; they are sometimes only a few inches in size, and composed merely of dark green confervæ, or purple or yellow lichens; but they are sometimes even of some feet in diameter, and contain seeds and various species of common water-plants, which are usually more or less incrusted with marble. There is, I believe, no place in the world where there is a more striking example of the opposition or contrast of the laws of animate and inanimate nature, of the forces of inorganic chemical affinity and those of the powers of life. Vegetables, in such a temperature, and every where surrounded by food, are produced with a wonderful rapidity; but the crystallizations are formed with equal quickness, and they are no sooner produced than they are destroyed together. The quantity of vegetable matter and its heat make it the resort of an infinite variety of insect tribes; and, even in the coldest days in winter, numbers of flies may be observed on the vegetables surrounding its banks or on its floating islands, and a quantity of their larvæ may be seen there, sometimes incrusted and entirely destroyed by calcareous matter, which is likewise often the fate of the insects themselves, as well as of various species of shell-fish that are found amongst the vegetables which grow and are destroyed in the travertine on its banks.

* * * * *

"I have passed many hours, I may say, many

days, in studying the phenomena of this wonderful lake; it has brought many trains of thought into my mind connected with the early changes of our globe, and I have sometimes reasoned from the forms of plants and animals preserved in marble in this warm source, to the grander depositions in the secondary rocks, where the zoophytes or coral insects have worked upon a grand scale, and where palms and vegetables now unknown are preserved with the remains of crocodiles, turtles, and gigantic extinct animals of the *Sauri* genus, and which appear to have belonged to a period when the whole globe possessed a much higher temperature.

* * * * *

"Then, from all we know, this lake, except in some change in its dimensions, continues nearly in the same state in which it was described seventeen hundred years ago by Pliny, and I have no doubt contains the same kinds of floating islands, the same plants, and the same insects. During the fifteen years that I have known it, it has appeared precisely identical in these respects; and yet it has the character of an accidental phenomenon depending upon subterraneous fire. How marvellous then are those laws by which even the humblest types of organic existence are preserved, though born amidst the sources of their destruction, and by which a species of immortality is given to generations, floating, as it were, like evanescent bubbles on a stream raised from the deepest caverns of the earth, and instantly losing what may be called its spirit in the atmosphere!"

From this interesting discourse on the formation

of Travertine, the conversation naturally turned to Geology; and I shall here again be compelled to give another copious extract, in order to show what were the latest opinions of Sir H. Davy upon this subject. If any doubt could exist as to the views here given being those entertained by the author, it is at once removed by his letter to Mr. Poole, in which, alluding to the work under review, he says, "*It contains the essence of my philosophical opinions.*"

"On the geological scheme of the early history of the globe, there are only analogies to guide us, which different minds may apply and interpret in different ways; but I will not trifle with a long preliminary discourse. Astronomical deductions and actual measures by triangulation prove that the globe is an oblate spheroid flattened at the poles; and this form, we know, by strict mathematical demonstrations, is precisely the one which a fluid body revolving round its axis and become solid at its surface by the slow dissipation of its heat or other causes, would assume. I suppose, therefore, that the globe, in the first state in which the imagination can venture to consider it, was a fluid mass with an immense atmosphere, revolving in space round the sun, and that by its cooling, a portion of its atmosphere was condensed in water which occupied a part of the surface. In this state, no forms of life, such as now belong to our system, could have inhabited it; and I suppose the crystalline rocks, or, as they are called by geologists, the *primary* rocks, which contain no vestiges of a former order of things, were the results of the first consolidation on its surface. Upon the farther cooling, the water which more or less had covered it, contracted; depositions took place, shell-fish, and coral insects of the first creation began their labours, and islands appeared in the midst of the ocean, raised from the deep by the productive energies of millions of zoophytes. These islands became covered with vegetables fitted to bear a high temperature, such as palms, and various species of plants similar to those which now exist in the hottest part of the world. And the submarine rocks or shores of these new formations of land became covered with aquatic vegetables, on which various species of shell-fish and common fishes found their nourishment. The fluids of the globe in cooling deposited a large quantity of the materials they held in solution, and these deposits agglutinating together the sand, the immense masses of coral rocks, and some of the remains of the shells and fishes found round the shores of the primitive lands, produced the first order of *secondary* rocks.

"As the temperature of the globe became lower, species of the oviparous reptiles were created to inhabit it; and the turtle, crocodile, and various gigantic animals of the *Sauri* kind, seem to have haunted the bays and waters of the primitive lands. But in this state of things there was no order of events similar to the present,—the crust of the globe was exceedingly slender, and the source of fire a small distance from the surface. In consequence of contraction in one part of the mass, cavities were opened, which caused the entrance of water, and immense volcanic explosions took place, raising one part of the surface, depressing another, producing mountains and causing new and extensive depositions from the primitive ocean. Changes of this kind must have been extremely frequent in the early epochas of nature; and the only living forms of which the remains are found in the strata that are the monuments of these changes, are those of plants, fishes, birds, and oviparous reptiles, which seem most fitted to exist in such a war of the elements.

"When these revolutions became less frequent, and the globe became still more cooled, and the inequalities of its temperature preserved by the mountain chains, more perfect animals became its inhabitants, many of which, such as the mammoth, megalonix, megatherium, and gigantic hyena, are now extinct. At this period, the temperature of the ocean seems to have been not much higher than it is at present, and the changes produced by occasional eruptions of it have left no consolidated rocks. Yet one of these eruptions appears to have been of great extent and of some duration, and seems to have been the cause of those immense quantities of water-worn stones, gravel, and sand, which are usually called *diluvian* remains; — and it is probable that this effect was connected with the elevation of a new continent in the southern hemisphere by volcanic fire. When the system of things became so permanent, that the tremendous revolutions depending upon the destruction of the equilibrium between the heating and cooling agencies were no longer to be dreaded, the creation of man took place; and since that period there has been little alteration in the physical circumstances of our globe.

Volcanoes sometimes occasion the rise of new islands, portions of the old continents are constantly washed by rivers into the sea, but these changes are too insignificant to affect the destinies of man, or the nature of the physical circumstances of things. On the hypothesis that I have adopted, however, it must be remembered, that the present surface of the globe is merely a thin crust surrounding a nucleus of fluid ignited matter; and consequently, we can hardly be considered as actually safe from the danger of a catastrophe by fire

* * * * *

"I beg you to consider the views I have been developing as merely hypothetical, one of the many resting-places that may be taken by the imagination in considering this subject. There are, however, distinct facts in favour of the idea, that the interior of the globe has a higher temperature than the surface; the heat increasing in mines the deeper we penetrate, and the number of warm sources which rise from great depths, in almost all countries, are certainly favourable to the idea. The opinion, that volcanoes are owing to this general and simple cause, is, I think, likewise more agreeable to the analogies of things, than to suppose them dependent upon partial chemical changes, such as the action of air and water upon the combustible bases of the earths and alkalies, though it is extremely probable that these substances may exist beneath the surface, and may occasion some results of volcanic fire;—and on this subject my notion may perhaps be the more trusted, as for a long while I thought volcanic eruptions were owing to chemical agencies of the newly discovered metals of the earths and alkalies, and I made many and some dangerous experiments in the hope of confirming this notion, but in vain.

* * * * *

"I have no objection to the '*refined Plutonic view*,' (of Professor Playfair and Sir James Hall,) as capable of explaining many existing phenomena; indeed, you must be aware that I have myself had recourse to it. What I contend against is, its application to explain the formations of the secondary rocks, which I think clearly belong to an order of facts not at all embraced by it. In the Plutonic system, there is one simple and constant order assumed, which may be supposed eternal. The surface is constantly imagined to be disintegrated, destroyed, degraded, and washed into the bosom of the ocean by water, and as constantly consolidated, elevated, and regenerated by fire; and the ruins of the old form the foundations of the new world. It is supposed that there are always the same types both of dead and living matter,—that the remains of rocks, of vegetables, and animals of one age are found imbedded in rocks raised from the bottom of the ocean in another. Now, to support this view, not only the remains of living beings which at present people the globe, might be expected to be found in the oldest secondary strata, but even those of the art of man, the most powerful and populous of its inhabitants, which is well known not to be the case. On the contrary, each stratum of the secondary rocks contains remains of peculiar and mostly now unknown species of vegetables and animals. In those strata which are deepest, and which must consequently be supposed to be the earliest deposited, forms even of vegetable life are rare; shells and vegetable remains are found in the next order; the bones of fishes and oviparous reptiles exist in the following class; the remains of birds, with those of the same genera mentioned before, in the next order; those of quadrupeds of extinct species in a still more recent class; and it is only in the loose and slightly consolidated strata of gravel and sand, and which are usually called diluvian formations, that the remains of animals, such as now people the globe, are found, with others belonging to extinct species. But in none of these formations, whether called secondary, tertial, or diluvial, have the remains of man, or any of his works, been discovered. It is, I think, impossible to consider the organic remains found in any of the earlier secondary strata, the lias-limestone and its congenerous formations, for instance, without being convinced, that the beings whose organs they formed belonged to an order of things entirely different from the present. Gigantic vegetables, more nearly allied to the palms of the equatorial countries than to any other plants, can only be imagined to have lived in a very high temperature; and the immense reptiles, the *Megalosauri*, with paddles instead of legs, and clothed in mail, in size equal, or even superior to the whale; and the great amphibia *Plethiosauri*, with bodies like turtles, but furnished with necks longer than their bodies, probably to enable them to feed on vegetables growing in the shallows of the primitive ocean, seem to show a state in which low lands, or extensive shores, rose above an immense calm sea, and when there were no great mountain chains to produce inequalities of temperature, tempests, or storms. Were the surface of the earth now to be carried down into the depths of the ocean, or were some great revolution of the waters to cover the existing land, and it was again to be elevated by fire, covered with consolidated depositions of sand or mud, how entirely different would it be in its characters from any of the secondary strata! Its great features would undoubtedly be the works of man: hewn stones, and statues of bronze and marble, and tools of iron, and human remains, would be more common than those of animals, on the greatest part of the surface; the columns of Pæstum, or of Agrigentum, or the immense iron bridges of the Thames, would offer a striking contrast to the bones of the crocodiles, or *Sauri*, in the older rocks, or even to those of the mammoth, or *Elephas primogenius*, in the diluvial strata. And whoever dwells upon this subject must be convinced, that the present order of things, and the comparatively recent existence of man, as the master of the globe, is as certain as the destruction of a former and a different order, and the extinction of a number of living forms, which have now no types in being, and which have left their remains wonderful monuments of the revolutions of nature."

The FOURTH DIALOGUE, to which is given the title of "The Proteus, or Immortality," is of a more desultory nature than those which precede it. It contains many beautiful descriptions of scenery in the Alpine country of Austria; furnishes an interesting account of that most singular reptile the

Proteus Anguinus, which is found only in the limestone caverns of Carniola, and concludes with reflections upon the indestructibility of the sentient principle.

The author's companion, during the tour he describes, is a scientific friend, whom he calls *Eubathes.* The dialogue opens with a passage of considerable pathos and eloquence: the author having been recalled to England by a melancholy event, the death of a very near and dear relation, describes his feelings on entering London.

"In my youth, and through the prime of manhood, I never entered London without feelings of pleasure and hope. It was to me as the grand theatre of intellectual activity, the field of every species of enterprise and exertion, the metropolis of the world of business, thought, and action. There, I was sure to find the friends and companions of my youth, to hear the voice of encouragement and praise. There, society of the most refined kind offered daily its banquets to the mind, with such variety that satiety had no place in them, and new objects of interest and ambition were constantly exciting attention either in politics, literature, or science.

"I now entered this great city in a very different tone of mind—one of settled melancholy, not merely produced by the mournful event which recalled me to my country, but owing likewise to an entire change in the condition of my physical, moral, and intellectual being. My health was gone, my ambition was satisfied; I was no longer excited by the desire of distinction; what I regarded most tenderly was in the grave; and to take a metaphor, derived from the change produced by time in the juice of the grape, my cup of life was no longer sparkling, sweet, and effervescent; it had lost its sweetness without losing its power, and it had become bitter."

There is perhaps not a more splendid passage to be found in the work; and it is scarcely inferior to Dr. Johnson's memorable conclusion to the preface of his Dictionary.

"After passing a few months in England," says he, "and enjoying (as much as I could enjoy any thing) the society of the few friends who still remained alive, the desire of travel again seized me. I had preserved amidst the wreck of time, one feeling strong and unbroken—the love of natural scenery; and this, in advanced life, formed a principal motive for my plans of conduct and action."

The fall of the Traun, about ten miles below Gmünden, was one of his favourite haunts; and he describes an accident of the most awful description which befell him at this place. While amusing himself on the water by a rapid species of locomotion, in a boat so secured by a rope as to allow only of a limited range, the tackle gave way, and he was rapidly precipitated down the cataract. He remained for some time after his rescue in a state of insensibility, and on recovering found himself attended by his mysterious friend the "Unknown," who had so charmed him in his excursion to Pæstum.

With this stranger, he proceeded on his tour; and he again becomes the medium through which much philosophical information is conveyed to the reader.

They visit together the grotto of the Maddalena at Adelsberg, and he gives us the conversation that took place in that extraordinary cavern.

"*Philalethes.*—If the awful chasms of dark masses of rock surrounding us appear like the work of demons, who might be imagined to have risen from the centre of the earth, the beautiful works of nature above our heads may be compared to a scenic representation of a temple or banquet-hall for fairies or genii, such as those fabled in the Arabian romances.

"*The Unknown.*—A poet might certainly place here the palace of the king of the Gnomes, and might find marks of his creative power in the small lake close by, on which the flame of the torch is now falling; for, there it is that I expect to find the extraordinary animals which have been so long the objects of my attention.

"*Eubathes.*—I see three or four creatures, like slender fish, moving on the mud below the water.

"*The Unknown.*—I see them; they are the Protei,—now I have them in my fishing-net, and now they are safe in the pitcher of water. At first view, you might suppose this animal to be a lizard, but it has the motions of a fish. Its head, and the lower part of its body and its tail, bear a strong resemblance to those of the eel; but it has no fins; and its curious bronchial organs are not like the gills of fishes; they form a singular vascular structure, as you see, almost like a crest, round the throat, which may be removed without occasioning the death of the animal, who is likewise furnished with lungs. With this double apparatus for supplying air to the blood, it can live either below or above the surface of the water. Its fore-feet resemble hands, but they have only three claws or fingers, are too feeble to be of use in grasping, or supporting the weight of the animal; the hinder feet have only two claws or toes, and in larger specimens are found so imperfect as to be almost obliterated. It has small points in place of eyes, as if to preserve the analogy of nature. Its nasal organs appear large; and it is abundantly furnished with teeth, from which it may be concluded that it is an animal of prey; yet in its confined state it has never been known to eat, and it has been kept alive for many years by occasionally changing the water in which it is placed.

"*Eubathes.*—Is this the only place in Carniola where these animals are found?

"*The Unknown.*—They were first discovered here by the late Baron Zois; but they have since been found, though rarely, at Sittich, about thirty miles distant, thrown up by water from a subterraneous cavity; and I have lately heard it reported that some individuals of the same species have been recognised in the calcareous strata in Sicily. I think it cannot be doubted, that their natural residence is an extensive deep subterranean lake, from which in great floods they sometimes are forced through the crevices of the rocks into this place where they are found; and it does not appear to me impossible, when the peculiar nature of the country in which we are is considered, that the same great cavity may furnish the individuals which have been found at Adelsberg and at Sittich.

* * * * *

"This adds one more instance to the number already known of the wonderful manner in which life is produced and perpetuated in every part of our globe, even in places which seem the least suited to organized existence. And the same infinite power and wisdom which has fitted the camel and the ostrich for the deserts of Africa, the swallow that secretes its own nest for the caves of Java, the whale for the Polar seas, and the morse and white bear for the Arctic ice, has given the Proteus to the deep and dark subterraneous lakes of Illyria,—an animal to whom the presence of light is not essential, and who can live indifferently in air and in water, on the surface of the rock, or in the depths of the mud."

Much interesting physiological discussion follows. I shall, however, merely notice the opinion delivered by the "Unknown," on the subject of respiration, and which I think shows that, at the conclusion of his career, Davy entertained the same notions, with regard to the communication of some ethereal principle to the blood, as he maintained in the earlier part of his life.*—"The obvious chemical alteration of the air is sufficiently simple in this process; a certain quantity of carbon only is added to it, and it receives an addition of heat or vapour; the volumes of elastic fluid inspired and expired (making allowance for change of temperature,) are the same, and if ponderable agents only were to be regarded, it would appear as if the only use of respiration were to free the blood from a certain quantity of carbonaceous matter. But it is probable that

* See vol. i. page 70.

this is only a secondary object, and that the change produced by respiration upon the blood is of a much more important kind. Oxygen, in its elastic state, has properties which are very characteristic; it gives out light by compression, which is not certainly known to be the case with any other elastic fluid except those which oxygen has entered without undergoing combustion; and from the fire it produces in certain processes, and from the manner in which it is separated by positive electricity in the gaseous state from its combinations, it is not easy to avoid the supposition, that it contains, besides its ponderable elements, some very subtile matter which is capable of assuming the form of heat and light. *My idea* is, that the common air inspired enters into the venous blood entire, in a state of dissolution, carrying with it its subtile or ethereal part, which in ordinary cases of chemical change is given off; that it expels from the blood carbonic acid gas and azote; and that, in the course of the circulation, its ethereal part and its ponderable part undergo changes which belong to laws that cannot be considered as chemical,—the ethereal part probably producing animal heat and other effects, and the ponderable part contributing to form carbonic acid and other products. The arterial blood is necessary to all the functions of life, and it is no less connected with the irritability of the muscles and the sensibility of the nerves than with the performance of all the secretions."

The FIFTH DIALOGUE is entitled "The Chemist." Its object is to demonstrate the importance of this noble science. An interlocutor is made to disparage its utility, and to mark its weaker points. These of course are answered, the sceptic becomes a true believer, and the intellectual gladiators separate mutually satisfied with each other.

" *Eubathes.*—I feel disposed to join you in attacking this favourite study of our friend, *but merely* to provoke him to defend it, in order to call forth his skill and awaken his eloquence.

" *The Unknown.*—I have no objection. Let there be a fair discussion: remember, we fight only with foils, and the point of mine shall be covered with velvet."

After having enumerated the scientific attainments necessary to constitute the chemist, and described the apparatus essential for understanding what has been already done in the science, he proceeds to define the intellectual qualities which he considers necessary for discovery, or for the advancement of the science. Amongst them, patience, industry, and neatness in manipulation, and accuracy and minuteness in observing and registering the phenomena which occur, are essential. A steady hand and a quick eye are most useful auxiliaries; but there have been very few great chemists who have preserved these advantages through life; for the business of the laboratory is often a service of danger, and the elements, like the refractory spirits of romance, though the obedient slaves of the magician, yet sometimes escape the influence of his talisman, and endanger his person. Both the hands and eyes of others, however, may be sometimes advantageously made use of. By often repeating a process or an observation, the errors connected with hasty operations or imperfect views are annihilated; and, provided the assistant has no preconceived notions of his own, and is ignorant of the object of his employer in making the experiment, his simple and bare detail of facts will often be the best foundation for an opinion. With respect to the higher qualities of intellect necessary for understanding and developing the general laws of the science, the same talents, I believe, are required as for making advancement in every other department of human knowledge; I need not be very minute. The imagination must be active and brilliant in seeking analogies; yet entirely under the influence of the judgment in applying them. The memory must be extensive and profound; rather, however, calling up general views of things than minute trains of thought;—the mind must not be like an Encyclopedia,—a burthen of knowledge, but rather a critical Dictionary, which abounds in generalities, and points out where more minute information may be obtained.

* * * * *

"In announcing even the greatest and most important discoveries, the true philosopher will communicate his details with modesty and reserve; he will rather be a useful servant of the public, bringing forth a light from under his cloak when it is needed in darkness, than a charlatan exhibiting fireworks, and having a trumpeter to announce their magnificence.

"I see you are smiling, and think what I am saying in bad taste; yet, notwithstanding, I will provoke your smiles still farther, by saying a word or two on his other moral qualities. That he should be humble-minded, you will readily allow, and a diligent searcher after truth, and neither diverted from this great object by the love of transient glory or temporary popularity, looking rather to the opinion of ages than to that of a day, and seeking to be remembered and named rather in the epochas of historians than in the columns of newspaper writers or journalists. He should resemble the modern geometricians in the greatness of his views and the profoundness of his researches, and the ancient alchemists in industry and piety. I do not mean that he should affix written prayers and inscriptions* of recommendations of his processes to Providence, as was the custom of Peter Wolfe, who was alive in my early days; but his mind should always be awake to devotional feelings; and in contemplating the variety and the beauty of the external world, and developing its scientific wonders, he will always refer to that Infinite Wisdom, through whose beneficence he is permitted to enjoy knowledge; and, in becoming wiser, he will become better,—he will rise at once in the scale of intellectual and moral existence, his increased sagacity will be subservient to a more exalted faith, and in proportion as the veil becomes thinner through which he sees the causes of things, he will admire more the brightness of the divine light by which they are rendered visible."

* In illustration of the pious custom here alluded to by Sir H. Davy, it may be observed, that the vessels of the alchemists very commonly bore some emblem; such, for instance, as that of the cross; and from which, indeed, the word *crucible* derived its appellation.

The SIXTH AND LAST DIALOGUE, entitled "POLA, or TIME," presents a series of reflections, to which a view of the decaying amphitheatre at Pola, an ancient town in the peninsula of Istria, is represented as having given origin. On former occasions, the inspection of the mouldering works of past ages called up trains of thought rather of a moral than of a physical character; in the present dialogue, the effects of time are considered in their relations to the mechanical and chemical laws by which material forms are destroyed, or rather changed,—for the author has shown by a number of beautiful examples, that without decay there can be no reproduction, and that the principle of change is a principle of life.

Having considered the influence of gravitation, the chemical and mechanical agencies of water, air, and electricity, and the energies of organized beings, in producing those diversified phenomena which, in our metaphysical abstractions, we universally refer to Time, he proceeds to enquire how far art can counteract their operation. A great philosopher, he observes, has said, "Man can in no other way command Nature but in obeying her laws:" it is evident that, by the application of some of those principles which she herself employs, we may for a while arrest the progress of changes which are ultimately inevitable.

"Yet, when all is done that can be done in the work of conservation, it is only producing a difference in the degree of duration. It is evident that none of the works of a mortal can be eternal, as none of the combinations of a limited intellect can be infinite. The operations of Nature, when slow, are no less sure; however man may, for a time, usurp dominion over her, she is certain of recovering her empire. He converts her rocks, her stones, her trees, into forms of palaces, houses, and ships; he employs the metals found in the bosom of the earth, as instruments of power, and the sands and clays which constitute its surface, as ornaments and resources of luxury; he imprisons air by water, and tortures water by fire to change, or modify, or destroy the natural forms of things. But in some lustrums his works begin to change, and in a few centuries they decay and are in ruins; and his mighty temples, framed as it were for immortal and divine purposes, and his bridges formed of granite and ribbed with iron, and his walls for defence, and the splendid monuments by which he has endeavoured to give eternity even to his perishable remains, are gradually destroyed; and these structures, which have resisted the waves of the ocean, the tempests of the sky, and the stroke of the lightning, shall yield to the operation of the dews of heaven,—of frost, rain, vapour, and imperceptible atmospheric influences; and as the worm devours the lineaments of his mortal beauty, so the lichens and the moss and the most insignificant plants shall feed upon his columns and his pyramids, and the most humble and insignificant insect shall undermine and sap the foundations of his colossal works, and make their habitations amongst the ruins of his palaces and the falling seats of his earthly glory."

On no occasion can such a subject be presented to a contemplative mind, without filling it with awe and wonder; but the circumstances under which these reflections are presented to us, in the last days of our philosopher, impress upon them an almost oracular solemnity. When we remember that while the mind of the philosopher was thus engaged in identifying the processes of decay with those of renovation in the system of nature, his body was palsied, and the current of his life fast ebbing, we cannot but admire that active intelligence which sparkled with such undiminished lustre amidst the wreck of its earthly tenement.

In the extracts which have been introduced from this last work, I trust the pledge that was given in the earlier part of these memoirs, has been redeemed by showing that a powerful imagination is not necessarily incompatible with a sound judgment, that the flowers of fancy are not always blighted by the cold realities of science, but that the poet and philosopher may, under the auspices of a happy genius, mutually assist each other in expounding the mysteries of nature. It cannot be denied, as a general aphorism, that the tree which expands its force in flowers is generally deficient in fruit; but the mind of Davy, to borrow one of his own metaphors, may be likened to those fabled of the Hesperides, which produced at once buds, leaves, blossoms, and fruits.

The happy effects resulting from this rare and nicely adjusted combination of talents, offer themselves as interesting subjects of biographical contemplation, and they can be studied only with success by a comparative analysis of different minds.

That the superiority of Davy greatly depended

upon the vivacity and compass of his imagination cannot be doubted, and such an opinion was well expressed by Mr. Davies Gilbert, in his late address to the Society :—" The poetic bent of Davy's mind seems never to have left him. To that circumstance I would ascribe the distinguishing features in his character and in his discoveries:—a vivid imagination sketching out new tracks in regions unexplored, for the judgment to select those leading to the recesses of abstract truth."

I have always thought that the mind of the late Dr. Clarke, the Mineralogical Professor of Cambridge, was little less imaginative than that of Davy; but it was deficient in judgment, and therefore often conducted him to error instead of to truth. Dr. Black was not deficient in imagination, and certainly not in judgment; but there was a constitutional apathy, arising probably from ill health, which damped his noblest efforts. *

* In addition to the anecdote already related of him, the following may serve to give a still greater force to this opinion. Soon after the appearance of Mr. Cavendish's paper on hydrogen gas, in which he made an approximation to the specific gravity of that body, showing that it was at least ten times lighter than common air, Dr. Black invited a party of his friends to supper, informing them that he had a curiosity to show them. Dr. Hutton and several others assembled, when, having the allentois of a calf filled with hydrogen gas, upon setting it at liberty, it immediately ascended, and adhered to the ceiling. The phenomenon was easily accounted for: it was taken for granted that a small black thread had been attached to the allentois,—that this thread passed through the ceiling, and that some one in the apartment above, by pulling the thread, elevated it to the ceiling, and kept it in that position. This explanation was so probable, that it was acceded to by the whole company; though, like many other plausible theo-

It is by the rarity with which the talent of seizing upon remote analogies is associated with a spirit of patient and subtile investigation of details, and a quick perception of their value, that the fact so truly stated by Mr. Babbage is to be explained; *viz.* that long intervals frequently elapse between the discovery of new principles in science and their practical application: thus he observes, that "the principle of the hydrostatic paradox was known as a speculative truth in the time of Stevinus, as far back as the year 1600,—and its application to raising heavy weights has long been stated in elementary treatises on natural philosophy, as well as constantly exhibited in lectures; yet it may fairly be regarded as a mere abstract principle, until the late Mr. Bramah, by substituting a pump, instead of the smaller column, converted it into a most valuable and powerful engine. The principle of the convertibility of the centres of oscillation and suspension in the pendulum, discovered by Huygens more than a century and a half ago, remained, until within these few years, a sterile though most elegant proposition; when, after being hinted at by Prony, and distinctly pointed out by Bonenberger, it was employed by Captain Kater as the foundation of a most conve-

ries, it was not true; for when the allentois was brought down, no thread whatever was found attached to it. Dr. Black explained the cause of the ascent to his admiring friends; but such was his unaccountable apathy, that he never gave the least account of this curious experiment even to his class; and more than twelve years elapsed before this obvious property of hydrogen gas was applied to the elevation of air balloons, by M. Charles, in Paris.

I am indebted for this anecdote to the "History of Chemistry," a very able work by Dr. Thomson, constituting the third number of the National Library.

nient method of determining the length of the pendulum. The interval which separated the discovery of Dr. Black, of latent heat, from the beautiful and successful application of it to the steam-engine, was comparatively short; but it required the efforts of two minds; and both were of the highest order."*

The discoveries of Davy present themselves in striking contrast with such instances. The same powerful genius that developed the laws of electro-chemical decomposition, was the first to apply them for the purpose of obviating metallic corrosion; and the nature of *fire-damp*, and the fact of its combustion being arrested in its passage through capillary tubes, were alike the discoveries of him who first applied them for the construction of a safety-lamp.†

In contrasting the genius of Wollaston with that of Davy, let me not be supposed to invite a comparison to the disparagement of either, but rather to the glory of both, for by mutual reflection each will glow the brighter. If the animating principle of

* "Reflections on the Decline of Science in England," page 15.

† While upon this subject, it is impossible not to notice the discoveries of Dr. Franklin, who combined in a remarkable degree a fertile imagination with a solid judgment; and the fruit of this union is to be seen in the invention of conductors for the security of ships and buildings against the effects of lightning. The philosopher who, predicating the identity of lightning and electricity, conceived the bold and grand idea of drawing it down from the thunder-cloud, an experiment which in another age would have consigned him to the dungeon for impiety, or to the stake for witchcraft, himself applied this wonderful discovery to the preservation of buildings, by the invention of pointed rods of iron. Of this invention it may be truly said, that he beat Nature with her own weapons, and triumphed over her power by an obedience to her own laws.

Davy's mind was a powerful imagination, generalizing phenomena, and casting them into new combinations, so may the striking characteristic of Wollaston's genius be said to have been an almost superhuman perception of minute detail. Davy was ever imagining something greater than he knew; Wollaston always knew something more than he acknowledged:—in Wollaston, the predominant principle was to avoid error; in Davy, it was the desire to discover truth. The tendency of Davy, on all occasions, was to raise probabilities into facts; while Wollaston as continually made them subservient to the expression of doubt.

Wollaston was deficient in imagination, and under no circumstances could he have become a Poet; nor was it to be expected that his investigations should have led him to any of those comprehensive generalizations which create new systems of philosophy. He well knew the compass of his powers, and he pursued the only method by which they could be rendered available in advancing knowledge. He was a giant in strength, but it was the strength of Antæus, mighty only on the earth. The extreme caution and reserve of his manner were inseparably connected with the habits of his mind; they pervaded every part of his character; in his amusements and in his scientific experiments, he displayed the same nice and punctilious observation,—whether he was angling for trout,* or testing for ele-

* Sir Humphry Davy has told us an anecdote which well illustrates this observation, while it affords a gratifying testimony of the kind feeling he entertained towards a kindred philosopher.— "There was—alas! that I must say *there was!*—an illustrious phi-

ments, he alike relied for success upon his subtile discrimination of minute circumstances.

By comparing the writings as well as the discoveries of these two great philosophers, we shall readily perceive the intellectual distinctions I have endeavoured to establish. "From their fruits shall ye know them." The discoveries of Davy were the results of extensive views and new analogies; those of Wollaston were derived from a more exact examination of minute and, to ordinary observers, scarcely appreciable differences. This is happily illustrated by a comparison of the means by which each discovered new metals. The alkaline bases were the products of a comprehensive investigation, which had developed a new order of principles; the detection of palladium and rhodium among the ores of platinum, was the reward of delicate manipulation, and microscopic scrutiny. As chemical operators, I have already pointed out their striking peculiarities, and they will be found to be in strict keeping with the other features of their respective characters. I might extend the parallel farther; but

losopher, who was nearly of the age of fifty before he made angling a pursuit, yet he became a distinguished fly-fisher, and the amusement occupied many of his leisure hours, during the last twelve years of his life. He indeed applied his preeminent acuteness, his science, and his philosophy, to aid the resources and exalt the pleasures of this amusement. I remember to have seen Dr. Wollaston, a few days after he had become a fly-fisher, carrying at his button-hole a piece of Indian rubber, when by passing his silkworm link through a fissure in the middle, he rendered it straight, and fit for immediate use. Many other anglers will remember other ingenious devices of my admirable and ever-to-be-lamented friend."—*Salmonia. Additional Note*, Edit. 2.

Dr. Henry, in the eleventh edition of his "System of Chemistry," has delineated the intellectual portraits of these two philosophers with so masterly a hand, that by quoting the passage, all farther observation will be rendered unnecessary.

"To those high gifts of nature, which are the characteristics of genius, and which constitute its very essence, both those eminent men united an unwearied industry and zeal in research, and habits of accurate reasoning, without which even the energies of genius are inadequate to the achievement of great scientific designs. With these excellencies, common to both, they were nevertheless distinguishable by marked intellectual peculiarities. Bold, ardent, and enthusiastic, Davy soared to greater heights; he commanded a wider horizon; and his keen vision penetrated to its utmost boundaries. His imagination, in the highest degree fertile and inventive, took a rapid and extensive range in pursuit of conjectural analogies, which he submitted to close and patient comparison with known facts, and tried by an appeal to ingenious and conclusive experiments. He was imbued with the spirit, and was a master in the practice, of the inductive logic; and he has left us some of the noblest examples of the efficacy of that great instrument of human reason in the discovery of truth. He applied it, not only to connect classes of facts of more limited extent and importance, but to develope great and comprehensive laws, which embrace phenomena that are almost universal to the natural world. In explaining those laws, he cast upon them the illumination of his own clear and vivid conceptions; he felt an intense admiration of the beauty, order, and harmony, which are conspicuous in the perfect Chemistry of Nature;—and he expressed those feelings with a force of eloquence which could issue only from a mind of the highest powers and of the finest sensibilities. With much less enthusiasm from temperament, Dr. Wollaston was endowed with bodily senses* of extraordinary acuteness and accuracy, and with great general vigour of understanding. Trained in the discipline of the exact sciences, he had acquired a powerful command over his attention, and had habituated himself to the most rigid correctness, both of thought and of language. He was sufficiently provided with the resources of the mathematics, to be enabled to pursue with success profound enquiries in mechanical and optical philosophy, the results of which enabled him to unfold the causes of phenomena not before understood, and to enrich the arts connected with those sciences, by the invention of ingenious and valuable instruments. In Chemistry, he was distinguished by the extreme nicety and delicacy of his observations; by the quickness and precision with which he marked resemblances and discriminated differences; the sagacity with which he devised experiments and anticipated their results; and

* Mr. Babbage considers it as a great mistake to suppose that Dr. Wollaston's microscopic accuracy depended upon the extraordinary acuteness of the bodily senses; a circumstance, he says, which, if it were true, would add but little to his philosophical character. He is inclined to view it in a far different light, and to see in it one of the natural results of the precision of his knowledge and of the admirable training of his intellectual faculties.

the skill with which he executed the analysis of fragments of new substances, often so minute as to be scarcely perceptible by ordinary eyes. He was remarkable, too, for the caution with which he advanced from facts to general conclusions: a caution which, if it sometimes prevented him from reaching at once to the most sublime truths, yet rendered every step of his ascent a secure station, from which it was easy to rise to higher and more enlarged inductions. Thus these illustrious men, though differing essentially in their natural powers and acquired habits, and moving, independently of each other, in different paths, contributed to accomplish the same great ends—the evolving new elements;- the combining matter into new forms; the increase of human happiness by the improvement of the arts of civilized life; and the establishment of general laws, that will serve to guide other philosophers onwards, through vast and unexplored regions of scientific discovery."

My history draws towards a conclusion.—Sir Humphry Davy, during the latter days of his life, was cheered by the society and affectionate attentions of his godson, the son of his old friend Mr. James Tobin.* He had been the companion of his travels, and he was the solace of his declining hours.

He had been resident for some months at Rome, where he occupied the second floor of a house in Via di Pietra, a street that leads out of the Corso.

* This inestimable man died on his plantation at Nevis, on the 19th of October 1814, in the forty-eighth year of his age.

During this period, he declined receiving any visitors, and had constantly some one by his side reading light works of interest to him, an amusement which was even continued during his meals.

As soon as the account of Sir Humphry having sustained another paralytic seizure was communicated to Lady Davy, who was in London at the time, she immediately set off, and so rapidly was her journey performed, that she reached Rome in little more than twelve days. Dr. John Davy, also, hastened from Malta, on receiving intelligence of his brother's imminent danger.

During his slow and partial recovery from this seizure, he learnt the circumstance of his name having been introduced into parliamentary proceedings, in the following manner. On the 26th of March 1829, on presenting a petition in favour of the Catholic claims from a very great and most respectable meeting at Edinburgh, Sir James Mackintosh, after having mentioned the name of Sir Walter Scott as being at the head of the petitioners, continued thus:—"Although not pertinent to this petition, yet connected with the cause, I indulge in the melancholy pleasure of adding to the first name in British literature the first name in British science — that of Sir Humphry Davy. Though on a sick-bed at Rome, he was not so absorbed by his sufferings as not to feel and express the glow of joy that shot across his heart at the glad tidings of the introduction of a bill which he hailed as alike honourable to his religion and his country."

I am assured that the last mark of satisfaction which he evinced from any intelligence communicated to him was on reading the above passage. He showed a pleasure unusual in his state of languor at the justice thus done, in the face of his country, to his consistency, to his zeal for religion and liberty, and to the generous sentiments which cheered his debility. The marks of his pleasure were observed by those who were brought most near to him by the performance of every kind office.

Although there appeared some faint indications of reviving power, his most sanguine friends scarcely ventured to indulge a hope that his life would be much longer protracted. Nor did he himself expect it. The expressions in his Will (printed in an Appendix) sufficiently testify the opinion he had for some time entertained of the hopelessness of his case.

In addition to this Will, he left a paper of directions, which have been religiously observed by his widow. He desires, for instance, that the interest arising from a hundred pounds stock may be annually paid to the Master of the Penzance Grammar School, on condition that the boys may have a holiday on his birthday.* There is something singularly interesting in this favourable recollection of his native town, and of the associations of his early youth. It adds one more example to show that, whatever may have been our destinies, and however fortune may have changed our conditions, where the heart remains uncorrupted, we shall, as the

* I understand that the present Master, the Reverend Mr. Morris, has expressed his intention to apply the above sum to purchasing a medal, which he intends to bestow as a Prize to the most meritorious scholar.

world closes upon us, fix our imaginations upon the simplicities of our youth, and be cheered and warmed by the remembrance of early pleasures, hallowed by feelings of regard for the memory of those who have long since slept in the grave.

With that restlessness which characterises the disease under which Sir Humphry Davy suffered, he became extremely desirous of quitting Rome, and of establishing himself at Geneva. His friends were naturally anxious to gratify every wish; and Lady Davy kindly preceded him on the journey, in order that she might at each stage make arrangements for his comfortable reception. Apartments were prepared for him at *L'Hotel de la Couronne*, in the Rue du Rhone; and at three o'clock on the 28th of May, having slept the preceding evening at Chambery, he arrived at Geneva, accompanied by his brother, Mr. Tobin, and his servant.

At four o'clock he dined, ate heartily, was unusually cheerful, and joked with the waiter about the cookery of the fish, which he appeared particularly to admire; and he desired that, as long as he remained at the hotel, he might be daily supplied with every possible variety that the lake afforded. He drank tea at eleven, and having directed that the feather-bed should be removed, retired to rest at twelve.

His servant, who slept in a bed parallel to his own, in the same alcove, was however very shortly called to attend him, and he desired that his brother might be summoned. I am informed that, on Dr. Davy's entering the room, he said, "I am dying;" or words to that effect; "and when it is all over, I desire that no disturbance of any kind may be made in the house; lock the door, and let every one retire quietly to his apartment." He expired at a quarter before three o'clock, without a struggle.

On the following morning, his friends Sismondi* and De Candolle were sent for; and the Syndics, as soon as the circumstance of his death was communicated to them, gave directions for a public funeral on the Monday; at which the magistrates, the professors, the English residents at Geneva, and such inhabitants as desired it, were invited to attend. The ceremony was ordered to be conducted after the custom of Geneva, which is always on foot—no hearse; nor did a single carriage attend. The cemetery is at Plain-Palais, some little distance out of the walls of the town. The Couronne being at the opposite extremity, the procession was long.

The following was the order of the procession:—†

The Two Syndics, (*in their robes*) { M. MASTOW, M. GALLATIN.

Magistrates of the Republic, { M. FAZIO, M. SALLADIN.

Professors of the College, in their robes,
MM. Simond de Sismondi—A. de Candolle.

THE ENGLISH.

Lord EGLINGTON,	Capt. ARCHIBALD HAMILTON,
Lord TWEDELL,	Mr. CAMPBELL,
Right Hon. WM. WYCKHAM,	Mr. FRANKS,

* Simond de Sismondi, the celebrated author of the History of the Italian Republic.

† For these particulars I am indebted to Sir Egerton Brydges.

Wm. Hamilton, Esq., Ex-Ambassador at Naples.	Mr. Alcock,
Sir Egerton Brydges, Bart.	Mr. Drew,
Colonel Alcock,	Mr. Heywood,
Captain Swinters,	Mr. Sitwell,
	&c.

The Students of the College.

The Citizens of Geneva.

The English service was performed by the Rev. John Magers, of Queen's College, and the Rev. Mr. Burgess.

The grave was stated in the public prints to be next to that of his friend the late Professor M. A. Pictet: this is not the fact. It is far away from it, on the second line of No. 29, the fourth grave from the end of the west side of the Cemetery.

Sir Humphry Davy having died without issue, his baronetcy has become extinct.

At present, the only memorial raised to commemorate the name of this distinguished philosopher is a Tablet placed in Westminster Abbey by his widow. It is thus inscribed :—

TO THE MEMORY OF
SIR HUMPHRY DAVY, BARONET;
DISTINGUISHED THROUGHOUT THE WORLD
BY HIS
DISCOVERIES IN CHEMICAL SCIENCE.
PRESIDENT OF THE ROYAL SOCIETY;
MEMBER OF THE NATIONAL INSTITUTE OF FRANCE.
BORN 17 DECEMBER 1778, AT PENZANCE.
DIED 28 MAY 1829, AT GENEVA,
WHERE HIS REMAINS ARE INTERRED.

The numerous scientific societies of which he was a member, will, no doubt, consecrate his memory. An eloquent Eloge has been read by Baron Cuvier before the Institute of France; but it has not yet been published: I have obtained, however, a copy of a speech delivered upon the same occasion, by H. C. Van der Boon Mesch, before the Institute of the Netherlands.

Mr. Davies Gilbert, his early friend and patron, has likewise paid to his memory a just and appropriate testimony of respect and admiration, in an address from the chair of the Royal Society.

The inhabitants of Penzance and its neighbourhood, animated by feelings of honourable pride and strong local attachment, will shortly, it is understood, raise a pyramid of massive granite to his memory, on one of those elevated spots of silence and solitude, where he delighted in his boyish days to commune with the elements, and where the Spirit of Nature moulded his genius in one of her wildest moods.

As yet, no intention on the part of the Government to commemorate this great philosopher, by the erection of a national monument, has been manifested: for the credit, however, of an age which is so continually distinguished as the most enlightened period in our history, I do hope the disgrace of such an omission may pass from us; although, I confess, it is rather to be wished than expected, when it is remembered that not a niche has been graced by the statue of Watt, while the giant iron children of his inventive genius are serving mankind in every quarter of the civilized world. A very erroneous impression would seem to exist with regard to the object and importance of such monuments. They are not to honour the dead, but to improve the living; not to give lustre to the philosopher, but to afford a salutary incentive to the disciple; not to perpetuate discoveries, for they can never be lost; but to animate scientific genius, and to engage it upon objects that may be useful to the commonwealth. Let it be remembered, that the ardour of the Roman youth was kindled into active emulation, whenever they beheld the images of their ancestors.

"Nam sæpe audivi, Q. Maximum, P. Scipionem, præterea civitatis nostræ præclaros viros, solitos ita dicere, cùm majorum imagines intuerentur, vehementissimè sibi animum ad virtutem accendi. Scilicet non ceram illam, neque figuram tantam vim in sese habere; sed memoriâ rerum gestarum eam flammam egregiis viris in pectore crescere, neque prius sedari, quàm virtus eorum famam atque gloriam adæquaverit."*

The fame of such a philosopher as Davy can never be exalted by any frail memorial which man can raise. His monument is in the great Temple of Nature.† His chroniclers are Time and the Elements. The destructive agents which reduce to dust the storied urn, the marble statue, and the towering pyramid, were the ministers of his power, and their work of decomposition is a perpetual memorial of his intelligence.

* Sallust. Bell. Jugurth.

† Ἀνδρῶν γὰρ ἐπιφανῶν πᾶσα γῆ τάφος, καὶ οὐ στηλῶν μόνον ἐν τῇ οἰκείᾳ σημαίνει ἐπιγραφὴ, ἀλλὰ καὶ ἐν τῇ μὴ προσηκούσῃ ἄγραφος μνήμη παρ᾽ ἑκάστῳ τῆς γνώμης μᾶλλον ἢ τοῦ ἔργου ἐνδιαιτᾶται.—Thucydides, B. 43.

A SKETCH OF THE HISTORY OF CHEMICAL SCIENCE, WITH A VIEW TO EXHIBIT THE REVOLUTIONS PRODUCED IN ITS DOCTRINES BY THE DISCOVERIES OF SIR HUMPHRY DAVY.

THE rapidity with which chemical opinions have risen into notice, flourished for a while, and then fallen into disrepute, to be succeeded by others equally precarious in their tenure and ephemeral in their popularity, are circumstances which the superficial reasoner has ever deplored, and the Sciolist as constantly converted into arguments against the soundness of the science which produced them. The leaves of a season will sprout, expand, and wither; and the dry foliage will be pushed off by the propulsion of new buds; but this last change is not effected in them, until they have absorbed the light and dews of heaven for the nourishment of the plant that bore them; and when even they shall have fallen to the earth, they will farther supply its spreading roots with fresh soil for its future growth and healthy developement; and entering into new combinations, will re-appear in the same tree under fresh forms of usefulness and symmetry. In like manner, chemical theories are but for a season; they are nothing more than general expressions of known facts; they may delight by their ingenuity, as vegetable forms captivate by their beauty, but their real and substantial use is to extend science; and as facts accumulate under their operation, they must give way to others better adapted to the increased growth and expansion of knowledge; nor does the utility of theories cease with their rejection,—they afford objects of analogy and comparison which assist the philosopher in his progress to truth, while their elements furnish materials for future arrangements. Were it otherwise, we should behold science in its advancement as a shapeless mass, enlarging by constant appositions, but without a single sign of growth or inward sympathy.

If chemical theories have undergone more rapid and frequent changes than those of other branches, the circumstance has arisen from the rapid manner in which new and important facts have been successively added to the general store.

Whatever may be the vices attributed to Chemistry on such occasions, they have belonged to the philosophers engaged in its pursuit, and are no evidence of the frailty of the science itself; and here it must be admitted, that there exists in one portion of mankind a self-love which cannot patiently submit to a change of opinions of which they are either the authors or defenders, while in another there predominates a timidity which naturally leads them, amidst the storm of controversy, to cling to the wreck of a shattered theory, rather than to trust themselves to a new and untried bark.

In our review of the history of Science, we have frequently to witness how the wisest philosopher has strained truth, for the support of a favourite doctrine, and measured and accommodated facts to theory, instead of adapting theories to facts—but this vice does not belong exclusively to chemical philosophers. Huygens, the celebrated Dutch Astronomer, from some imaginary property in the number *six*, having discovered *one* of Saturn's moons, absolutely declined looking for any more, merely because that one, when added to the four moons of Jupiter, and to the one belonging to the Earth, made up the required number.

Such reflections naturally arise on viewing, with a philosophic eye, the progress and modifications of chemical opinions; and it is essential that they should be duly appreciated upon the present occasion; for, before any just estimate can be formed of the talents and services of Sir Humphry Davy, we must thoroughly consider, in all their bearings and relations, the various prejudices with which he had to contend in his efforts to modify a gigantic theory, which enjoyed an unrestrained dominion in the chemical world, and for many years continued to be the pride of France and the admiration of Europe.

It would be quite foreign to the plan of this sketch,* which the reader must consider as wholly subservient to the object that has been announced, to enquire how far the ancients, in their metallurgical processes, can be said to have exercised the arts of chemistry. Equally vain would it be to enter into a history of that system of delusion and imposture, so long practised under the denomination of Alchymy. It is only necessary to consider Chemistry in its

* This historical sketch has no pretensions to originality. It is compiled from the best authors, and from the Introduction to Sir H. Davy's Elements of Chemical Philosophy.

dignified and purely scientific form; and we have only to notice those commanding discoveries and opinions which led to the developement of that system, which the genius of Davy was destined to modify.

The origin of Chemistry, as a science, cannot be dated farther back than about the middle of the seventeenth century; and Beccher, the contemporary of Boyle, who was born at Spires in 1635, was unquestionably the first to construct any thing like a general theory. He formed the bold idea of explaining the whole system of the earth by the mutual agency and changes of a few elements. And by supposing the existence of a vitrifiable, a metallic, and an inflammable earth, he attempted to account for the various productions of rocks, crystalline bodies, and metallic veins, assuming a continual interchange of principles between the atmosphere, the ocean, and the solid surface of the globe, and considering the operations of nature as all capable of being imitated by art.

Albertus Magnus had advanced the opinion that the metals were earthy substances impregnated with a certain inflammable principle; but Beccher supported the idea of this principle not only as the cause of metallization, but likewise of combustibility. Stahl, however, one of the most extraordinary men that Germany ever produced, having adopted and amplified this theory, carried off the entire credit of being its founder, and it is universally spoken of as the *Stahlian Theory*.

This theory forms so important a feature in the history of chemistry, and so long maintained its ascendency in the schools, that it will be necessary to give the reader a short summary of its principles. It assumed that all *combustible* bodies are compounds: one of the constituents being volatile, and therefore easily dissipated during the act of combustion; while the other, being fixed, constantly remained as the residue of the process. This volatile principle, for which Stahl invented the term *Phlogiston*, was considered as being identical in every species of combustible matter; in short, it was supposed that there was but one principle of combustibility in nature, and that was the imaginary phantom Phlogiston, which for nearly a century possessed the schools of Europe, and, like an evil spirit, crossed the path of the philosopher at every step, and by its treacherous glare allured him from the steady pursuit of truth; for, whether a substance were combustible or not, its nature could never be investigated without a reference to its supposed relations with Phlogiston; its presence, or its absence, was supposed to stamp a character upon all bodies, and to occasion all the changes which they undergo. Hence chemistry and combustion came to be in some measure identified; and a theory of combustion was considered the same thing as a theory of chemistry.

The identity of Phlogiston in all combustible bodies was founded upon observations and experiments of so decisive a nature, that after the existence of the principle itself was admitted, they could not fail to be satisfactory. When phosphorus is made to burn, it gives out a strong flame, much heat is evolved, and the phosphorus is dissipated in fumes, which, if properly collected, will quickly absorb moisture from the atmosphere, and produce an acid liquid known by the name of phosphoric acid. Phosphorus then must consist, say the Stahlians, of Phlogiston and this acid. Again—If this liquid be evaporated to a dry substance, mixed with a quantity of charcoal powder, and then heated in a vessel from which the external air is excluded, a *portion*, or the *whole* of the charcoal will disappear, and phosphorus will be reproduced, possessing all the properties that it had before it was subjected to combustion. In this case, it was supposed that the charcoal restored the phlogiston. There was much plausibility in all this, as well as in the reasoning which followed. Since we may employ, with equal success, any kind of combustible body for the purpose of changing phosphoric acid into phosphorus, such as lamp-black, sugar, resin, or even several of the metals, it was concluded that all such bodies contain a common principle which they communicate to the phosphoric acid; and since the new body formed is in all cases identical, the principle communicated must also be identical. Hence combustible bodies contain an identical principle, and this principle is Phlogiston.

The same theory applied with equal force to the burning of sulphur and several of the metals, and to their reconversion by combustible bodies.

When lead is kept nearly at a red-heat in the open air for some time, it is converted into a pigment called *red lead;* this is a calx of lead. To restore this calx again to metallic lead, it is only necessary to heat it in contact with almost any combustible matter; all these bodies therefore must contain one common principle, which they communicated to the red lead, and by so doing reconverted it to the state of metal. Metals then were regarded as compounds of *calces* and phlogiston. Thus far the theory works glibly enough; but now comes a startling fact, which was long unnoticed by the blind adherents of Stahl, or, if noticed, intentionally overlooked. It was observed very early, that when a metal was converted into a calx, its weight was increased. When this difficulty first forced itself upon the attention of the Phlogistians, it was necessary that they should either explain it, or at once abandon their theory. They accordingly endeavoured to evade the difficulty, not only by asserting that phlogiston had no weight, but that it was actually endowed with a principle of levity.

It was not possible, however, that any rational notions should have been entertained upon the subject of combustion, at a period when the composition of the atmosphere even was unknown. Let us therefore follow the stream of discovery, skimming the surface merely, as it flowed onward towards quite a new field of science—Pneumatic Chemistry.

Boyle and Hooke, who had improved the air-pump invented by Otto de Guericke, of Madenburgh, first used this apparatus for investigating the properties of air; and they concluded from their experiments that air was absolutely necessary to combustion and respiration, and that one part of it only was employed in these processes; and Hooke formed the sagacious conclusion, that this principle is the same as the substance fixed in nitre, and that combustion is a chemical process, the solution of the burning body in elastic fluid, or its union with this matter.

Mayow, of Oxford, in 1674, published his treatises on the Nitro-aërial spirit, in which he advanced opinions similar to those of Boyle and Hooke, and supported them by a number of original and curious experiments.

Dr. Hales, about 1724, resumed the investigations commenced with so much success by Boyle, Hooke, and Mayow; and endeavoured to ascertain the chemical relations of air to other substances, and to ascertain by statistical experiments the cases in nature, in which it is absorbed or emitted. He obtained a number of curious and important results; he disengaged elastic fluids from various substances, and drew the conclusion, that air was a chemical element in many compound bodies, and that flame resulted from the action and reaction of aërial and sulphurous particles; but all his reasonings were contaminated with the notion of one elementary principle constituting elastic matter, and modified in its properties by the effluvia of solid or fluid bodies.

The light of Pneumatic science which had dawned under Hooke, Mayow, and Hales, burst forth in splendour under the ascendency of that constellation of British science, Black, Cavendish, and Priestley.

In 1756, Dr. Black published his researches on calcareous, magnesian, and alkaline substances, by which he proved the existence of a gaseous body, perfectly distinct from the air of the atmosphere. He showed, that quick-lime differed from marble and chalk by not containing this substance, which he proved to be a weak acid, capable of being expelled from alkaline and earthy bodies by stronger acids.

As nothing is more instructive than to enquire into the circumstances which have led to a great discovery, I quote with pleasure the following passage from Dr. Thomson's History of Chemistry.

"It was the good fortune of chemical science that, at this time (1751), the opinions of professors were divided concerning the manner in which certain lithonthriptic medicines, particularly lime-water, acted in alleviating the excruciating pains of the stone and gravel. The students usually partake of such differences of opinion: they are thereby animated to more serious study, and science gains by their emulation.

"All the medicines which were then in vogue as solvents of calculi had a greater or less resemblance to caustic potash or soda; substances so acrid, when in a concentrated state, that in a short time they reduce the fleshy parts of the animal body to a mere pulp. They all seemed to derive their efficacy from quick-lime, which again derived its power from the fire. It was therefore very natural for them to ascribe its power to igneous matter imbibed from the fire, retained by the lime, and communicated by it to alkalies which it renders powerfully acrid. It appears from Dr. Black's note-books, that he originally entertained the opinion, that caustic alkalies acquired igneous matter from quick-lime. In one of them, he hints at some way of catching this matter as it escapes from lime, while it becomes mild by exposure to the air; but on the opposite blank page is written, 'Nothing escapes—the cup rises considerably by absorbing air.' A few pages further on, he compares the loss of weight sustained by an ounce of chalk when calcined, with its loss while dissolved in muriatic acid.

"These experiments laid open the whole mystery, as appears by another memorandum. 'When I precipitate lime by a common alkali, there is no effervescence: the air quits the alkali for the lime; but it is lime no longer, but C. C. C: it now effervesces, which good lime will not.'—What a multitude of important consequences naturally flowed from this discovery! He now knew to what the causticity of alkalies is owing, and how to induce it, or remove it, at pleasure. The common notion was entirely reversed. Lime imparts nothing to the alkalies; it only removes from them a peculiar kind of air *(carbonic acid gas)* with which they were combined, and which prevented their natural caustic properties from being developed. All the former mysteries disappear, and the greatest simplicity appears in those operations of nature which before appeared so intricate and obscure."

Dr. Thomson afterwards observes,—"The discovery which Dr. Black had made, that marble is a combination of lime and a peculiar substance, to which he gave the name of *fixed air*, began gradually to attract the attention of chemists in other parts of the world. It was natural, in the first place, to examine the nature and properties of this fixed air, and the circumstances under which it is generated. It may seem strange and unaccountable that

Dr. Black did not enter with ardour into this new career which he had himself opened, and that he allowed others to reap the corn after having himself sown the grain. Yet he did take some steps towards ascertaining the properties of *fixed air;* though I am not certain what progress he made. He knew that a candle would not burn in it, and that it is destructive to life, when any living animal attempts to breathe it. He knew that it is formed in the lungs during the breathing of animals, and that it is generated during the fermentation of wine and beer. Whether he was aware that it possesses the properties of an acid, I do not know; though with the knowledge which he possessed that it combines with alkalies and alkaline earths, and neutralizes them, or at least blunts and diminishes their alkaline properties, the conclusion that it partook of acid properties was scarcely avoidable. All these, and probably some other properties of *fixed air*, he was in the constant habit of stating in his lectures from the very commencement of his academical career; though, as he never published any thing on the subject himself, it is not possible to know exactly how far his knowledge of the properties of *fixed air* extended. The oldest manuscript copy of his lectures that I have seen was taken down in writing in the year 1773; and before that time Mr. Cavendish had published his paper on *fixed air* and *hydrogen gas*, and had detailed the properties of each. It was impossible from the manuscript of Dr. Black's lectures, to know which of the properties of *fixed air* stated by him were discovered by himself, and which were taken from Mr. Cavendish."

An idea so novel and important as that of an air possessing properties quite different from that of the atmosphere, existing in a fixed and solid state in various bodies, was not received without doubt, and even opposition. Several German enquirers endeavoured to controvert it. Meyer attempted to show that limestone became caustic, not by the emission of elastic matter, but by combining with a peculiar substance in the fire; the loss of weight, however, was wholly inconsistent with such a view of the question: and Bergman at Upsal, Macbride in Ireland, Keir at Birmingham, and Cavendish in London, fully demonstrated the truth of the opinion of Black, and a few years were sufficient to establish his theory upon an immutable foundation, and to open a new road to most important discoveries.

The knowledge of one elastic fluid, entirely different in its properties from air, very naturally suggested the probability of the existence of others. The processes of fermentation which had been observed by the ancient chemists, and those by which Hales had disengaged and collected elastic substances, were now regarded under a novel point of view; and the consequence was, that a number of new bodies, possessed of very extraordinary properties, were discovered.

Mr. Cavendish, about the year 1765, invented an apparatus for examining elastic fluids confined by water, which has since been called the *hydro-pneumatic* apparatus. He discovered inflammable air, and described its properties; he ascertained the relative weights of fixed air, inflammable air, and common air, and made a number of beautiful and accurate experiments on the properties of these elastic substances.

Dr. Priestley, in 1771, entered the same path of enquiry; and principally by repeating the processes of Hales, added a number of most important facts to this department of chemical philosophy. He discovered nitrous air, nitrous oxide, and dephlogisticated air, (oxygen) and by substituting mercury for water in the pneumatic apparatus, ascertained the existence of several aëriform bodies which are rapidly absorbable by water; such as muriatic acid gas, sulphurous acid gas, and ammonia.

Scheele, independently of Priestley, also discovered several of the aëriform bodies; he ascertained likewise the composition of the atmosphere; he brought to light fluoric acid, prussic acid, and the substance which he termed *dephlogisticated marine acid*, the oxymuriatic acid of the French school, and the chlorine of Davy.

Sir Humphry Davy, in the preface to his Chemical Philosophy, observes that Black, Cavendish, Priestley, and Scheele, were undoubtedly the greatest chemical discoverers of the eighteenth century; and that their merits are distinct, peculiar, and of the most exalted kind. He thus defines them:

" Black made a smaller number of original experiments than either of the other philosophers; but being the first labourer in this new department of the science, he had greater difficulties to overcome. His methods are distinguished for their simplicity; his reasonings are admirable for their precision; and his modest, clear, and unaffected manner is well calculated to impress upon the mind a conviction of the accuracy of his processes, and the truth and candour of his researches.

"CAVENDISH was possessed of a minute knowledge of most of the departments of Natural Philosophy: he carried into his chemical researches a delicacy and precision, which have never been exceeded: possessing depth and extent of mathematical knowledge, he reasoned with the caution of a geometer upon the results of his experiments; and it may be said of him, what, perhaps, can scarcely be said of any other person, that whatever he accomplished, was perfect at the moment of its production. His processes were all of a finished nature; executed by the hand of a master, they required no correction; the accuracy and beauty of his earliest labours even have remained unimpaired amidst the progress of discovery, and their merits have been illustrated by discussion and exalted by time.

"DR. PRIESTLEY began his career of discovery without any general knowledge of chemistry, and with a very imperfect apparatus. His characteristics were ardent zeal and the most unwearied industry. He exposed all the substances he could procure to chemical agencies, and brought forward his results as they occurred, without attempting logical method or scientific arrangement. His hypotheses were usually founded upon a few loose analogies; but he changed them with facility; and being framed without much effort, they were relinquished with little regret. He possessed in the highest degree ingenuousness and the love of truth. His manipulations, though never very refined, were always simple, and often ingenious. Chemistry owes to him some of her most important instruments of research, and many of her most useful combinations; and no single person ever discovered so many new and curious substances.

"SCHEELE possessed in the highest degree the faculty of invention; all his labours were instituted with an object in view, and after happy or bold analogies. He owed little to fortune or to accidental circumstances: born in an obscure situation, occupied in the duties of an irksome employment, nothing could damp the ardour of his mind, or chill the fire of his genius; with very small means, he accomplished very great things. No difficulties deterred him from submitting his ideas to the test of experiment. Occasionally misled in his views, in consequence of the imperfection of his apparatus, or the infant state of the enquiry, he never hesitated to give up his opinions the moment they were contradicted by facts. He was eminently endowed with that candour which is characteristic of great minds, and which induces them to rejoice as well in the detection of their own errors, as in the discovery of truth. His papers are admirable models of the manner in which experimental research ought to be pursued; and they contain details on some of the most important and brilliant phenomena of chemical philosophy."

The discovery of the gases, of a new class of bodies more active than any others in most of the phenomena of nature and art, could not fail to modify the whole theory of chemistry, and, under the genius of Lavoisier, it ultimately led to the establishment of those new doctrines, which it is the principal object of this history to expound; but before this task can be accomplished, it will be necessary to consider the rise and progress of opinion concerning chemical attraction, and heat and light, since these subjects are too intimately interwoven with the *anti-phlogistic* system to be separated from any examination of its principles.

Boyle, says Sir Humphry Davy, was one of the most active experimenters, and certainly the greatest chemist of his age. He introduced the use of *tests,* or *re-agents,* active substances for detecting the presence of other bodies: he overturned the ideas which at that time were prevalent, that the results of operations by fire were the real elements of things; and he ascertained a number of important facts respecting inflammable bodies, and alkalies, and the phenomena of combination; but neither he nor any of his contemporaries endeavoured to account for the changes of bodies by any fixed principles.

The solutions of the phenomena were attempted either on rude mechanical notions, or by occult qualities, or peculiar subtile spirits or ethers, supposed to exist in the different bodies. And it is to the same great genius who developed the laws that regulate the motions of the heavenly bodies, that chemistry owes the first distinct philosophical elucidations of the powers which produce the changes and apparent transmutations of the substances belonging to the earth.

"Sugar dissolves in water, alkalies unite with acids, metals dissolve in acids. Is not this," says Newton, "on account of an attraction between their particles? Copper dissolved in aqua fortis is thrown down by iron. Is not this because the particles of the iron have a stronger attraction for the particles of the acid, than those of copper; and do not different bodies attract each other with different degrees of force?"

In 1719, Geoffroy endeavoured to ascertain the relative attractive powers of bodies for each other, and to arrange them, under the form of a table, in an order in which these forces, which he named affinities, were expressed.

Concerning the nature of heat, there are two opinions which have ever divided the chemical world. The one considers it merely as a property of matter, and that it consists in an undefinable motion, or vibration of its particles; the other, on the contrary, regards it as a distinct and subtile substance, *sui generis*. Each of these opinions has been supported by the greatest philosophers, and for a long period the arguments on both sides appeared equally plausible and forcible. The discovery of Dr. Black, however, gave a preponderance to the scale in favour of its materiality.

"It was during his residence in Glasgow, between the year 1759 and 1763," says Dr. Thomson, "that he brought to maturity those speculations concerning the combination of *heat* with *matter*, which had frequently occupied a portion of his thoughts."

Before Dr. Black's discovery, it was universally supposed that solids were converted into liquids by a small addition of heat, after they have been once raised to the melting point, and that they returned again to the solid state on a very small diminution of the quantity of heat necessary to keep them at that temperature. An attentive view, however, of the phenomena of liquefaction and solidification gradually led this sagacious philosopher to a different conclusion. By observations which it is unnecessary to detail, he became satisfied that when ice is converted into water, it unites with a quantity of heat, without having its temperature increased; and that when water is frozen into ice, it gives out a quantity of heat without having it diminished. The heat thus combined, then, is the cause of the fluidity of the water; and as it is not sensible to the thermometer, Dr. Black called it *latent heat.*

There is such an analogy between the cessation of thermometric expansion during the liquefaction of ice, and during the conversion of water into steam, that there could be no hesitation about explaining both in the same way. Dr. Black, therefore, immediately concluded that, as water is ice united to a certain quantity of *latent* heat, so steam is water united to a still greater quantity.

This beautiful theory enables us to understand phenomena in nature which were previously quite inexplicable. We now comprehend how the thaw which supervenes after intense frost, should so slowly melt the wreaths of snow and beds of ice. Had, indeed, the transition of water from its solid into its liquid state not been accompanied by this great change in its relation to heat, every thaw would have occasioned a frightful inundation, and a single night's frost must have solidified our rivers and lakes. Neither animal nor vegetable life could have subsisted under such sudden and violent transitions. It would appear, then, that water, during the act of freezing, is acted upon by two opposite powers: it is deprived of heat by exposure to a medium whose temperature is below 32°; and it is supplied with heat by the evolution of that principle from itself, *viz.* of that portion which constituted its fluidity. As these powers are exactly equal, the temperature of the water must remain unchanged till the latent heat, necessary to its fluidity, is all evolved.

Although these facts have been admitted by all, it has been contended by many that the absorption of heat by bodies is the necessary *effect*, and not the efficient *cause*, of change of form, — the consequence of what has been called a change of their *capacity:* thus ice, it is supposed, in becoming water, has its capacity for heat increased, and the absorption of heat is a consequence of such increased capacity. This theory, however, is deficient, inasmuch as it fails to explain the cause of that change of form, which is assumed to account for the increase of capacity.

Light, like heat, has been considered by some philosophers as a subtile fluid filling space, and rendering bodies visible by the undulations into which it is thrown; while others, with Newton at their head, regard it as a substance consisting of small particles, constantly separating from luminous bodies, moving in straight lines, and rendering objects visible by passing from them and entering the eye. The late experiments of Dr. Young would incline us to prefer the undulatory to the corpuscular hypothesis.

By this preliminary sketch, the reader has been prepared for viewing with advantage the theory of Lavoisier; in the construction of which he will see little more than a happy generalization of the several discoveries which have been enumerated. Indeed, this observation will apply to all great systems of philosophy; facts, developed by successive enquirers, go on accumulating, until, after an interval, a happy genius arises who connects and links them together; and thus generally receives that meed of praise which, in stricter justice, would be apportioned and awarded to the separate contributors. It is far from my intention to disparage the merits of Lavoisier; but the materials of his system were undoubtedly furnished by Black, Priestley, and Cavendish.

The most important modification of the phlogistic theory—for there were several others—may be said to be that suggested by Dr. Crawford. Dr. Priestley had found that the air in which combustibles were suffered to burn till they were extinguished, underwent a very remarkable change, for no combustible would afterwards burn in it, and no animal could breathe it without suffocation. Dr. Crawford, like many others, concluded, that this change was owing to phlogiston; but he for the first time applied Dr. Black's doctrine of *latent* heat, for the explanation of the origin of the heat and light which appear during the process. According to this philosopher, the phlogiston of the combustible combines, during combustion, with the air, and at the same time separates the caloric and light with which that fluid had been previously united. The heat and the light, then, which appear during combustion, exist previously in the air. This theory was very different from Stahl's, and certainly a great deal more satisfactory; but still the question—*What is phlogiston?* remained to be answered.

Mr. Kirwan attempted to answer it, and to prove that phlogiston is no other than hydrogen.

This opinion, which Mr. Kirwan informs us was first suggested by the discoveries of Dr. Priestley, met with a very favourable reception from the chemical world, and was adopted, amongst many others, by Mr. Cavendish. The object of Mr. Kirwan was to prove, that hydrogen exists as a component part of every combustible body; that during combustion it separates from the combustible body, and combines with the oxygen of the air. At the same time, Lavoisier was engaged in examining the experiment of Bayen, and those of the British philosophers. Bayen, in 1774, had shown that mercury converted into a calx, or earth, by the absorption of air, could be revived without the addition of any inflammable substance; and hence he concluded, that there was no necessity for supposing the existence of any peculiar principle of inflammability, in order to account for the calcination of metals; but he formed no opinion respecting the nature of the air produced from the *calx* of mercury. Lavoisier, in 1775, showed that it was an air, which supported flame and respiration better than common air, which he afterwards named oxygen: the same substance that Priestley and Scheele had procured from other metallic bodies the year before, and had particularly described.

Lavoisier also discovered that the same air is produced during the revivification of metallic calces by charcoal, as that which is emitted during the calcination of limestones; hence he concluded, that this elastic fluid is composed of oxygen and charcoal: and from his experiments on nitrous acid and oil of vitriol, he also inferred that this gas entered into the composition of these substances.

Lavoisier was now enabled to explain the phenomena of combustion, without having recourse to phlogiston: a principle merely supposed to exist, because combustion could not be explained without it.

His new theory depends upon the two laws discovered by himself and Dr. Black; *viz.* that when a combustible is raised to a certain temperature, it begins to combine with the oxygen of the atmosphere, and that this oxygen during its condensation lets go the *latent* caloric, and the light with which it was combined while in the gaseous state. Hence their appearance during every combustion. Hence also the change which the combustible undergoes in consequence of combustion.

It followed from this view, that the metallic *calces* were combinations of metals with oxygen; and on examining the products of certain inflammable bodies, and finding them to be acid, the conclusion was extended by a plausible analogy to other acids whose bases were unknown, and the general proposition was established that oxygen was the universal principle of acidity; that acids resulted from the union of a peculiar combustible base, called the *radical*, with the common principle, oxygen, technically termed the *acidifier*.

These views, regarding the phenomena of combustion and acidification, may be considered as constituting what has been termed the *Anti-phlogistic system.*

It was some time, however, after this system was promulgated, before its author was able to gain a single convert, notwithstanding his unwearied assiduity, and the great weight which his talents, his reputation, his fortune, and his situation naturally gave him.

At length, M. Berthollet, at a meeting of the Academy of Sciences in 1785, solemnly renounced his old opinions, and declared himself a convert. Fourcroy followed his example; and two years afterwards Morveau, during a visit to Paris, was prevailed upon to embrace the new doctrine.

The theory of Lavoisier, soon after it had been framed, received an important confirmation from the two grand discoveries of Mr. Cavendish, respecting the composition of water and nitric acid, and the elaborate and beautiful investigations of Berthollet into the nature of ammonia; by which, phenomena, before anomalous, were shown to depend upon combinations of aëriform matter.

The notion of phlogiston, however, was still defended with remarkable tenacity by many distinguished philosophers. Mr. Kirwan, who considered hydrogen as the universal principle of combustibility, undertook to prove that this element entered into the composition of every body of the kind: a single exception, of course, must necessarily prove fatal to the theory. Mr. Kirwan, fortunately for the French chemists, founded his reasonings on the inaccurate experiments of other chemists; and thus did he promote the popularity of the anti-phlogistic system by the weakness of the arguments by which he assailed it.

Lavoisier and his associates saw at once the important uses which might be made of this essay: by refuting an hypothesis which had been embraced by the most respectable chemists in Europe, their cause would receive an *éclat* which would make it irresistible. The essay was accordingly translated into French, and each of the sections into which it was divided was accompanied by a refutation.

Four of the sections were refuted by Lavoisier, three by Berthollet, three by Fourcroy, two by Morveau, and one by Monge.

Mr. Cavendish, in a paper communicated to the Royal Society in the year 1784, drew a comparison between the phlogistic and anti-phlogistic theories, and showed that each of them was capable of explaining the phenomena in a satisfactory manner; he however, at the same time, gave the reasons which induced him to prefer the earlier view. In the execution of this task, unlike Mr. Kirwan, he never advanced a single opinion which he had not put to the test of experiment; and he never suffered himself to go any farther than his experiments would warrant. This paper, therefore, the French chemists were unable to refute, and they were accordingly wise enough to pass it over without notice. Had it been possible to have preserved the phlogistic hypothesis, Mr. Cavendish would have saved it—

"Si Pergama dextrâ
Defendi possent, etiam hac defensa fuissent."

"Sooner or later," says Sir Humphry Davy, "that doctrine which is an expression of facts, must prevail over that which is an expression of opinion. The most important part of the theory of Lavoisier was merely an arrangement of the facts relating to the combinations of oxygen: the principle of reasoning which the French school professed to adopt was, that every body which was not yet decompounded, should be considered as simple; and though mistakes were made with respect to the results of experiments on the nature of bodies, yet this logical and truly philosophical principle was not violated; and the systematic manner in which it was enforced was of the greatest use in promoting the progress of science."

Till 1786, there had been no attempt to reform the nomenclature of chemistry; the names applied by discoverers to the substances which they made known were still employed. Some of these names, which originated amongst the alchymists, were of the most barbarous kind; few of them were sufficiently definite or precise, and most of them were founded upon loose analogies, or upon false theoretical views.

"It was felt by many philosophers, particularly by the illustrious Bergman, that an improvement in chemical nomenclature was necessary; and in 1787, MM. Lavoisier, Morveau, Berthollet, and Fourcroy, presented to the world a plan for an almost entire change in the denomination of chemical substances, founded upon the idea of calling simple bodies by some names characteristic of their most striking qualities, and of naming compound bodies from the elements which composed them." There was, besides, a secret feeling in the breasts of the associated chemists, which, no doubt, had its influence in suggesting and promoting such a scheme. The views of Lavoisier had so changed the face of chemistry, as almost to have rendered it a new science: by adopting a new nomenclature, they identified, as it were, all the discoveries of the day with the new theory, and thus appropriated to France the original and entire merit of the system.

It is impossible to pass over this subject without a comment. Lavoisier was unquestionably indebted to Dr. Black for the support, if not for the suggestion, of the most brilliant part of his theory of combustion; and yet he attempted even to conceal the name of the discoverer of *latent heat.*

How far Lavoisier was really culpable, and whether he did not intend to do full justice to all the claims of his predecessors, cannot now be known; as he was cut off in the midst of his career, while so many of his scientific projects remained unexecuted. From the posthumous works of Lavoisier, there is some reason for believing that, if he had lived, he would have done justice to all parties; but there is no doubt that Dr. Black, in the mean time, thought himself aggrieved by the publication of several of Lavoisier's papers in the "Mémoires de l'Académie," and that he formed the intention of doing himself justice, by publishing an account of his own discoveries: this intention, however, was unfortunately thwarted and prevented by bad health. But to return to the subject of nomenclature. Sir H. Davy continues—" The new nomenclature was speedily adopted in France; under some modifications, it was received in Germany; and, after much discussion and opposition, it became the language of a new and rising generation of chemists in England. It materially assisted the diffusion of the anti-phlogistic doctrine, and even facilitated the general acquisition of the science; and many of its details were contrived with much address, and were worthy of its celebrated authors."

On the general adoption of this new theory of chemistry, it must be admitted that its authors displayed an intemperate triumph wholly unworthy of them. They held a festival, at which Madame Lavoisier, in the habit of a priestess, burnt the works of Stahl on an altar erected for the occasion, while solemn music played a requiem to his departed system!

Sir Humphry Davy, in speaking of the merits of Lavoisier, observes that "he must be regarded as one of the most sagacious of the chemical philosophers of the last century; indeed, except Cavendish, there is no other enquirer who can be compared to him for precision of logic, extent of view, and sagacity of induction. His discoveries are few, but he reasoned with extraordinary correctness upon the labours of others. He introduced weight and measure, and strict accuracy of manipulation into all chemical processes. His mind was unbiassed by prejudice; his combinations were of the most philosophical nature; and in his investigations upon ponderable substances, he has entered the true path of experiment with cautious steps, following just analogies, and measuring hypotheses by their simple relations to facts."

It will be scarcely possible for a future generation of philosophers to imagine with what an undisciplined ardour the anti-phlogistic system, thus enhanced by a new and fascinating nomenclature, was supported throughout Europe. Facts only were appreciated in proportion to the evidence they furnished of its truth; and a discovery even required the sanction of its authority as the passport to notice and regard. The least expression of doubt, as to the validity of any point in its doctrines, exposed the sceptic to a host of assailants, and fortunate was he if he escaped the fate of Peter Ramus, or of those who ventured to question the infallibility of that great despot of another age, Aristotle.

In no country of Europe did this feeling manifest itself to a greater extent than in England. There was perhaps a political prejudice co-operating upon the occasion: it is very difficult, under any circumstances, to avoid connecting the man and his works. The fate of Lavoisier* was truly affecting,

* Lavoisier perished on the scaffold at the age of fifty-one, during the sanguinary reign of Robespierre. The fury of the revolutionary leaders of France was particularly directed against the farmers-general of the revenue, who were all executed, with the exception of a single individual, a M. de Verdun. Sixty of them were guillotined at the same time, in consequence of a report of Dupin, a frantic member of the Convention. The revolutionary tribunal adopted a general formula, as the ground of their condemnation, which is curious as applied to Lavoisier, who was declared guilty of having "adulterated snuff with water and ingredients destructive of the health of the citizens." The unfortunate philosopher requested time to complete some experiments on respiration. The reply of Coffinhal, the President, was, that "the Republic did not want savans or chemists, and that the course of justice could not be suspended."

and by a species of retributive justice, he received the sympathy of the world in the homage paid to his system; while the atrocity of his assassination, on which every Englishman dwelt with horror, appeared to be thus heightened by every praise bestowed upon his merits.

It is not the least surprising circumstance in the history of this system, that with such a blind and idolatrous admiration of its principles, so few facts should have been distorted. It is true that, from the belief that combustion could never take place without the presence of oxygen, the elementary principle of Scheele became, according to these views, a compound of oxygen and an acid; and the name of *dephlogisticated marine acid* was exchanged for that of *oxymuriatic acid*, a circumstance which spread a cloud of error over the science, and perhaps retarded its progress in a greater degree than is generally imagined. In like manner, the chemist neglected to avail himself of the hint which, under other impressions, would have proved an important clue to discovery, *viz.* the acid properties of sulphuretted hydrogen.

We have now arrived at that stage in our history, when it may with propriety and advantage be asked —WHAT HAS DAVY DONE IN CORRECTING ERROR, OR IN ADVANCING TRUTH?

The answer to this question will be nothing more than a summary of those discoveries which have been successively investigated during the progress of the present work.

The new doctrines of chemistry were highly instrumental in encouraging more extended investigations into all the different productions of nature and art; and we may observe, that one of the first efforts of Sir Humphry Davy was to improve our knowledge of the nature and habitudes of the tanning and astringent principles of vegetables,—an enquiry which had been commenced by Seguin and Proust. In pursuing even the most beaten path, he was sure to discover objects of novelty. Look at his early experiments on the cane, and on the straw of wheat, barley, and hay, and we shall see how magically he raised from their ashes a new flower of knowledge. He soon, however, quitted the track of other experimentalists; although we learn from the whole tenor of his researches, that he could obey as well as he could command, and he could act in the ranks, although he more frequently appeared as a general in the field of science.

Sir Humphry Davy has observed, that "at the time when the anti-phlogistic theory was established, electricity had little or no relation to chemistry. The grand results of Franklin respecting the cause of lightning, had led many philosophers to conjecture, that certain chemical changes in the atmosphere might be connected with electrical phenomena; and electrical discharges had been employed by Cavendish, Priestley, and Van-Marum, for decomposing and igniting bodies; but it was not till the era of the wonderful discovery of Volta, in 1800, of a new electrical apparatus, that any great progress was made in chemical investigation by means of electrical combinations.

"Nothing tends so much to the advancement of knowledge as the application of a new instrument.

The native intellectual powers of men in different times are not so much the causes of the different success of their labours, as the peculiar nature of the means and artificial resources in their possession. Independent of vessels of glass, there could have been no accurate manipulations in common chemistry: the air-pump was necessary for the investigation of the properties of gaseous matter; and without the Voltaic apparatus, there was no possibility of examining the relations of electrical polarities to chemical attractions."

There is a candour in this statement which we cannot but admire. Nor does the admission diminish the glory of him who, by the application of such new instruments of research, was enabled to penetrate into the hidden mysteries of Nature. What avails the telescope, without the eye of the observer?

To Davy, the Voltaic apparatus was the *golden branch*, by which he subdued the spirits that had opposed the advance of former philosophers; but what would its possession have availed him, had not his genius, like the ancient Sibyl, pointed out its use and application?

It will be seen that he was thus enabled, not only to discover laws which are in constant operation, modifying the forms of matter, and influencing all the operations of chemistry, but, by applying them, to determine the elements of the fixed alkalies to be oxygen and a metallic base: a fact obviously opposed to the idea of oxygen being the general principle of acidity; for here it was the principle of alkalinity, if it may be so expressed. This was shaking the corner-stone of the edifice, and his subsequent researches into the nature of oxymuriatic acid may be said to have overthrown it; for if either of the elements of this body can be considered as the acidifier, it is hydrogen. The consequences which flowed from this truth were of the highest importance, not only in correcting errors, which the progress of discovery, instead of rectifying, was actually multiplying, but in leading to the developement of new bodies. Iodine might have been recognised as an elementary body; but its relations to oxygen and hydrogen would probably have remained unknown, had not a knowledge of the true character of chlorine assisted the enquiry.

The same observation will apply to the recently discovered body, Bromine. In like manner has the chemist been led, by the chloridic theory, to a more accurate acquaintance with the composition of the fluoric, hydriodic, and hydrocyanic acids; while he has also learnt that hydrogen alone can convert certain undecompounded bases into well characterised acids, without the aid of oxygen. The same discovery has completely changed all our opinions with regard to a very important series of saline combinations, and developed the existence of new compounds of a most interesting description.

Thus, then, has the *acidifying* hypothesis of Lavoisier been overturned, and a new theory constructed out of its ruins, which acknowledges no distinct element as the one imparting to matter the characters of an acid.

Equally complete has been the downfal of the theory of combustion. The discovery of the true nature of chlorine was, in itself, sufficient to show that bodies might combine, with the phenomena of heat and light, without the presence of oxygen ; but Davy has brought a mass of evidence from other sources in proof of the same truth. He has shown that, whenever the chemical forces which determine either composition or decomposition are energetically exercised, the phenomena of combustion, or incandescence, with a change of properties, are displayed. He has therefore annulled the distinction between supporters of combustion and combustibles, since he has shown that, in fact, one substance frequently acts in both capacities, being a supporter *apparently* at one time, and a combustible at another. But in both cases the heat and light depend on the same cause, and merely indicate the energy and rapidity with which reciprocal attractions are exerted. Thus sulphuretted hydrogen is a combustible with oxygen and chlorine; a supporter with potassium. Sulphur, with chlorine and oxygen, has been called a combustible basis; with metals, it acts the part of a supporter. In like manner, potassium unites so powerfully with arsenic and tellurium, as to produce the phenomena of combustion. Nor can we ascribe the appearances to the liberation of latent heat, in consequence of condensation of volume. The protoxide of chlorine, a body destitute of any combustible constituent, at the instant of decomposition evolves light and heat with explosive violence; and its volume becomes one-fifth greater. Chloride and iodide of azote, compounds alike destitute of any inflammable matter, according to the ordinary belief, are resolved into their respective elements with tremendous force of inflammation; and the first expands into more than six hundred times its bulk. Now, instead of heat and light, a prodigious degree of *cold* ought to accompany such an expansion, according to the hypothesis of latent heat. Other instances might be cited, and other arguments adduced on the same subject, but time and space fail me.*

Such, then, are the facts developed by the experimental researches of Sir Humphry Davy; from which it follows, that—

1. Combustion is not necessarily dependent on the agency of oxygen.

2. That it cannot be regarded as dependent upon any peculiar principle or form of matter, but must be considered as a general result of intense chemical action.

3. That the evolution of light and heat cannot be ascribed simply to a gas parting with its latent store of those ethereal fluids.

4. That, since all bodies which act powerfully upon each other are in the opposite electrical relations of positive and negative, the evolution of heat and light may depend upon the annihilation of these opposite states, which will happen whenever they combine.

Thus has Sir H. Davy, by refuting the opinions of the French philosophers, respecting the relations of oxygen to the phenomena of combustion, and the

* The reader who wishes for further details will consult with advantage the article Combustion in Dr. Ure's Dictionary of Chemistry; a work to which I acknowledge myself much indebted on this and other occasions.

nature of its products, removed the pillars on which the fabric of the anti-phlogistic rested, and reduced the generalization of Lavoisier to isolated collections of facts; the sound logic, however,—the pure candour, the numerical precision of inference which characterise the labours of the French philosopher, will cause his name to be held in everlasting admiration. The downfal of his doctrine is the natural result of the progress of truth; the same fate may attend our present systems, but the facts discovered through their means are unchangeable and eternal; and it is upon them alone that the fame of the chemist must ultimately rest.

In sciences collateral to chemistry, the researches of Davy have cast a reflected lustre. In geology, his discovery of the composition of the earths, has opened a new path of investigation; while his examination of the water and gaseous matter so frequently enclosed in the cavities of quartz, has given no small degree of support to the hypothesis of the Plutonists; above all, his results connected with the decomposition and transfer of different elements by Voltaic influence, has already explained many phenomena relating to metallic veins; and the late researches of Mr. Fox must lead us to the conclusion, that electric powers are still in operation in the recesses of the earth; and that mineral veins are not only the cabinets of Nature, but still her active laboratories.

These cursory observations upon the discoveries of Sir H. Davy relate merely to the changes they have effected in the general theory of chemistry. I might recapitulate the numerous researches by which he has extended our knowledge upon particular subjects; but I have so fully entered into the consideration of them in the body of my work, that I consider such a tax upon the patience of my reader would be both unfair and unnecessary

I shall therefore *conclude* my long and arduous labour, by enumerating the different memoirs communicated by this distinguished philosopher to the Royal Society; and also the several works which he published at different periods of his brilliant but too fleeting career.

1. An Account of some Galvanic Combinations, formed by single metallic plates and fluids, analogous to the Galvanic Apparatus of M. Volta. *Read June* 18, 1801.
2. An Account of some Experiments and Observations on the constituent parts of certain Astringent Vegetables, and on their operation in Tanning. *February* 24, 1803.
3. An Account of some Analytical Experiments on a Mineral Production from Devonshire, consisting principally of Alumina and Water. *February* 28, 1805.
4. On a Method of analysing Stones containing a fixed Alkali, by means of Boracic acid. *May* 16, 1805.

* *For the above papers, the Society awarded him the Copley Medal.*

5. The Bakerian Lecture.—On some Chemical agencies of Electricity. *November* 20, 1806.

** *For this memoir, he received the prize of the French Institute.*

6. The Bakerian Lecture.—On some new Phenomena of Chemical Changes produced by Electricity, particularly the Decomposition of the Fixed Alkalies, and the exhibition of the new substances which constitute their bases; and on the general nature of Alkaline bodies. *Read November* 19, 1807.

7. Electro-chemical Researches on the Decomposition of the Earths; with Observations on the Metals obtained from the Alkaline Earths; and on the Amalgam procured from Ammonia. *June* 30, 1808.

8. The Bakerian Lecture.—An Account of some new Analytical Researches on the nature of certain bodies, particularly the Alkalies, Phosphorus, Sulphur, Carbonaceous matter, and the Acids hitherto uncompounded; with some general Observations on Chemical Theory. *December* 15, 1808.

9. New Analytical Researches on the nature of certain bodies; being an Appendix to the Bakerian Lecture for 1808. *February* 1809.

10. The Bakerian Lecture for 1809. On some new Electro-chemical Researches on various objects, particularly the metallic bodies from the Alkalies and Earths; and on some combinations of Hydrogen. *November* 16, 1809.

11. Researches on the Oxy-muriatic Acid, its nature and combinations; and on the elements of Muriatic Acid; with some Experiments on

Sulphur and Phosphorus, made in the Laboratory of the Royal Institution.
Read July 12, 1810.

12. The Bakerian Lecture for 1810. On some of the Combinations of Oxy-muriatic Gas and Oxygen, and on the chemical relations of those principles to inflammable bodies.
November 15, 1810.

13. On a Combination of Oxy-muriatic Gas and Oxygen Gas.
February 21, 1811.

14. On some Combinations of Phosphorus and Sulphur, and on some other subjects of Chemical Enquiry. *June* 18, 1812.

15. On a new Detonating Compound; in a letter to Sir Joseph Banks, Bart. F.R.S.
November 5, 1812.

16. Some further Observations on a new Detonating substance. *July* 1, 1813.

17. Some Experiments and Observations on the Substances produced in different chemical processes on Fluor Spar.
July 8, 1813.

18. An Account of some New Experiments on the Fluoric Compounds; with some Observations on other objects of Chemical Enquiry.
February 13, 1814.

19. Some Experiments and Observations on a new Substance, which becomes a Violet-coloured Gas by heat.
January 20, 1814.

20. Further Experiments and Observations on Iodine. *Read June* 16, 1814.

21. Some Experiments on the Combustion of the Diamond, and other Carbonaceous Substances. *June* 23, 1814.

22. Some Experiments and Observations on the Colours used in Painting by the Ancients. *February* 23, 1815.

23. Some Experiments on a solid Compound of Iodine and Oxygen, and on its Chemical Agencies. *April* 20, 1815.

24. On the Action of Acid upon the Salts usually called *Hyper-Oxymuriates*, and on the Gases produced from them. *May* 4, 1815.

25. On the *Fire-damp* of Coal Mines, and on methods of lighting the Mine, so as to prevent Explosion. *November* 19, 1815.

26. An Account of an Invention for giving Light in Explosive Mixtures of *Fire-damp* in Coal Mines, by consuming the Fire-damp. *January* 11, 1816.

27. Further Experiments on the Combustion of Explosive Mixtures confined by Wire Gauze, with some Observations on Flame. *January* 25, 1816.

28. Some Researches on Flame. *January* 16, 1817

29. Some New Experiments and Observations on the Combustion of Gaseous Mixtures; with

an account of a method of preserving a continued Light in Mixtures of inflammable Gases and Air, without Flame.

Read January 23, 1817.

* * * *For the preceding five papers, the Rumford Medals were awarded to him.*

30. On the fallacy of Experiments in which Water is said to have been formed by the decomposition of Chlorine.

February 12, 1818.

31. New Experiments on some of the combinations of Phosphorus.

April 9, 1818.

32. Some Observations on the formation of Mists in particular situations.

February 25, 1819.

33. On the Magnetic Phenomena produced by Electricity.

November 16, 1820.

34. Some Observations and Experiments on the Papyri found in the ruins of Herculaneum.

March 15, 1821.

35. Further Researches on the Magnetic Phenomena produced by Electricity; with some new Experiments on the properties of Electrified bodies, in their relations to conducting powers and temperature.

July 5, 1821.

36. On the Electrical Phenomena exhibited *in vacuo.*

December 20, 1821.

37. On the state of Water and Aëriform matter in cavities found in certain Crystals.
Read June 13, 1822.

38. On a new Phenomenon of Electro-Magnetism.
March 6, 1823.

39. On the application of Liquids formed by the condensation of Gases, as mechanical Agents.
April 17, 1823.

40. On the changes of volume produced in Gases, in different states of density, by Heat.
May 1, 1823.

41. On the Corrosion of Copper Sheathing by sea-water; and on methods of preventing this effect, and on their application to ships of war and other ships. *Jan.* 24, 1824.

42. Additional Experiments and Observations on the application of Electrical Combinations to the preservation of the Copper Sheathing of ships, and to other purposes.
June 17, 1824.

43. Further Researches on the preservation of Metals by Electro-chemical means.
June 9, 1825.

44. The Bakerian Lecture for 1826.—On the Relation of Electrical and Chemical changes.
June 3, 1826.

**** *For this memoir, the Royal Society conferred upon him the Royal Medal.*

45. On the Phenomena of Volcanoes.
March 20, 1828.

46. Account of some Experiments on the Torpedo.
November 20, 1828.

HIS PUBLISHED WORKS ARE,

"Experimental Essays on Heat, Light, and on the Combinations of Light, with a new Theory of Respiration," &c. Published in *Contributions to Physical and Medical Knowledge, by T. Beddoes, M. D.* 1799.

"Researches Chemical and Philosophical, chiefly concerning Nitrous Oxide, and its Respiration." 1800.

"A Syllabus of a Course of Lectures."

"An Introductory Lecture." 1801.

"Elements of Chemical Philosophy." 1812.

"Elements of Agricultural Chemistry." 1813.

"On the Safety Lamp for Coal Miners; with some Researches on Flame." 1818. (Several Editions.)

"Salmonia; or Days of Fly-Fishing."

"Consolations in Travel; or the Last Days of a Philosopher."

APPENDIX.

APPENDIX.

EXTRACTED FROM THE REGISTRY OF THE PREROGATIVE COURT OF CANTERBURY.

A.

MY WILL.

This 3rd of January 1827 feeling more than usual symptoms of mortality I make this my Will. First, I give my Brother John Davy M.D. three hundred pounds a-year of money that I possess in the Long Annuities and likewise four thousand pounds to be raised by the sale of Securities I possess in the English or French funds or annuities but I mean my said Brother to devote the interest of three thousand pound of these last moneys to such purposes as he may deem fitting for the benefit of my sisters particularly my married one and I wish a part of the interest of these three thousand pounds to be employed in educating and settling in life my godson Humphrey Millett. I leave him Dr. Davy likewise all the property devolving to me from my parents which has never been divided to do what seems to him best for the benefit of my sisters and my sister Millett's children and I leave my said brother my Chemical Books and Chemical MSS. Apparatus *Sporting tackle* Medals and the silver Venetian dish made from the Rumford Medal in token of my affection. I leave £100 to each of these friends Dr. Babington and Dr. Franck and £50 to Dr. Wilson Philip and to Mr. Brodie surgeon to lay out in tokens of remembrance. I leave all my other property whether in goods money chattells funded securities annuities or plate to my wife (Lady) Jane Davy and I appoint her the sole Executrix of this my Will. If my brother or his family should not be in a condition at the time of her decease to use my service of plate given for the safety lamp I wish it to be sold and the same given to the Royal Society to provide an annual medal from the interest for the best discovery made any where in chemistry and I depend upon my dear wife to make such presents in seals or token to such of my friends as she may think proper agreeably to their and her feelings.

H. DAVY.

B.

Further explanatory Clause.

I leave to my wife Dame Jane Davy all my other property whether funded or in government securities or in leases of houses or goods &c. and I leave her my sole residuary legatee and sole Executrix. I wish her to enjoy the use of my plate during her life and that she will leave it to my brother in case he survive her and if not to any child of his who may be capable of using it but if he be not in a situation to use or enjoy it then I wish it to be melted and given to the Royal Society to found a medal to be given annually for the most important discovery in Chemistry any where made in Europe or Anglo-America. Knowing the perfect understanding and love of justice of my wife I leave to her all other arrangements which may make my memory useful to the world and awaken the kind feelings of my friends and I wish her and my brother and all my friends every happiness this life can afford.

HUMPHRY DAVY.

C.

That is a Clause explanatory of my Will.

I wish seals not rings with a fish engraved upon them to be given to some of my friends amongst whom I mention Mr Knight Dr Babington Mr Pepys Mr Hatchett. And lest there should be any doubt respecting the £3000 mentioned I mean my brother to be a trustee for this and should he die without children I mean it to belong to my sister Millett's children £2000 to Humphry Millett my godson and the rest to be equally divided between the other children but should my brother marry and have children I then mean after the death of my sisters these £3000 to be divided between her child or children and my sisters and £1000 to go to Humphry Millett my godson and £500 to my sister's other children leaving the arrangement to my brother.

H. D.

D.

Further explanatory Clause, Feb. 27th 1828.

I leave to my brother John Davy M.D. the proceeds of my Agricultural Chemistry in the future editions and the profits of my work on fishing and I give him the copyright. I leave my friend Thomas Poole Esq. of Stowey fifty pounds to purchase some token of remembrance.

H. D.

Rome Nov. 18th, 1828.

By this addition to my will I confirm all that I have willed in a paper left in a brass box at Messrs. Drummond leaving Lady Davy my sole Executrix and residuary legatee. I leave the copyright of Salmonia to my brother John Davy wishing him to apply a part of the profits of the sale of the editions of this work to the education of my nephew Humphrey Millett in case he has no children of his own. I leave the copy of my Vision in my writing desk to Lady Davy to be published if my friends think it may give pleasure or information to the public but I wish the profits of this work to be applied to the use of my brothers and sisters. I leave to Josephine Detela daughter of Mr. Detela of Laybach in Illyria innkeeper my kind and affectionate nurse one hundred pounds or rather a sum which shall equal a thousand florins to be paid out of the balance at my banker's within three months after my decease. I beg Lady Davy to be so good as to fulfill my engagements with the persons who are travelling with me but without any favour as I have no reason to praise either their attention or civilities within the last two months but the kindness and attentions of Josephine Detela during my illness at Laybach not only calls for the testimony I have given but likewise my gratitude for which I give her the £100 or the 1000 florins.

H. Davy.

Feb. 19th 1829.

I wish to be buried where I die *natura curat suas reliquias.* I wish £100 to be given to George Whidby and I beg Lady Davy to fulfill all my engagements and that if my friends should think my Dialogues worthy of publication I beg that they may be published and that Mr. Tobin may correct the press of them and I wish that £150 may be given to him for this labour. There is a codicil to my will in my writing desk. I beg Lady Davy to have the goodness to attend to every thing mentioned in that. In addition to what I have mentioned in that codicil I request that £50 or 500 florins may be given to Josephine Dettela within five months after my decease and I wish £50 to be presented to my friend Dr Morichini in remembrance and memory of his great kindness to me.

H. D.

I wish one hundred to be given to my amanuensis.

For the purpose of explaining a Will that I made before I left England and some papers that I have since added to it I write these few words Rome, March 18, 1829.

I give the copyright of Salmonia my Dialogues and any other of my works which my friends may think it proper to republish to my brother John Davy M.D. to be published in the manner he may think most fit and proper. I have already in my former testament left Lady Davy my residuary legatee but I beg her in considering the disposition of my property to regard £6000 as belonging to my brother Dr. Davy in case there rests any doubt upon this subject in my first will and I wish her the said Lady Davy to enjoy during her life the use and property of the different services of plate given to me whether by the Emperor of Russia or the different coal *committees* but I trust to her sense of justice that she will leave them in the manner I have pointed out in my will to my brother. With respect to any property at present in my banker's hands or any thing I now carry with me I leave them entirely to my brother Dr. Davy.

HUMPHRY DAVY.

At Rome March 18, 1829.

THE END.

LONDON:
PRINTED BY SAMUEL BENTLEY,
Dorset Street, Fleet Street.

Printed in the United States
By Bookmasters